The 10th International Seminar of Veterinary Medicine: Camelids in Algeria & Maghreb

The 10th International Seminar of Veterinary Medicine: Camelids in Algeria & Maghreb

Editors

Amira Leila Dib
Said Boukhechem
Hithem Bougherara
El-Hacene Bererhi

MDPI • Basel • Beijing • Wuhan • Barcelona • Belgrade • Manchester • Tokyo • Cluj • Tianjin

Editors

Amira Leila Dib
Institute of Veterinary Sciences, Université Frères Mentouri Constantine 1, Constantine, Algeria

Said Boukhechem
Institute of Veterinary Sciences, Université Frères Mentouri Constantine 1, Constantine, Algeria

Hithem Bougherara
Institute of Veterinary Sciences, Université Frères Mentouri Constantine 1, Constantine, Algeria

El-Hacene Bererhi
Institute of Veterinary Sciences, Université Frères Mentouri Constantine 1, Constantine, Algeria

Editorial Office
MDPI
St. Alban-Anlage 66
4052 Basel, Switzerland

This is a reprint of Proceedings from the conference volume published online in the open access journal *Biology and Life Sciences Forum* (ISSN 2673-9976) (available at: https://www.mdpi.com/2673-9976/22/1).

For citation purposes, cite each article independently as indicated on the article page online and as indicated below:

LastName, A.A.; LastName, B.B.; LastName, C.C. Article Title. *Journal Name* **Year**, *Volume Number*, Page Range.

ISBN 978-3-0365-7480-6 (Hbk)
ISBN 978-3-0365-7481-3 (PDF)

Cover image courtesy of Institute of Veterinary sciences, Constantine, Algeria.

Contents

biology and life sciences forum

MDPI

Editorial

Preface of the 10th International Seminar of Veterinary Medicine: Camelids in Algeria and the Maghreb

Amira Leila Dib *, Said Boukhechem, Hithem Bougherara and El-Hacene Bererhi

Gestion Santé et Productions Animales Research Laboratory, Institut des Sciences Vétérinaires El- Khroub, Université Frères Mentouri Constantine 1, Constantine 25000, Algeria; said.boukhechem@umc.edu.dz (S.B.); heithem1985@yahoo.fr (H.B.); brerhihacene@yahoo.fr (E.-H.B.)

* Correspondence: dib.amiraleila@umc.edu.dz

This book collects articles from the contributions presented at the 10th International Seminar of Veterinary Medicine: Camelids in Algeria and the Maghreb, held in Constantine, Algeria, at Université Frères Mentouri, Constantine 1, Institute of Veterinary Sciences, on 20–21 December 2022.

The history of the domestication of camelids remains poorly documented. Accordingly, in recent years, the dromedary has been the subject of particular research attention from the national and international authorities. This animal is the subject of capital importance and topicality in line with the 17 Sustainable Development Goals (SDGs), established by the Member States of the United Nations and collated in the 2030 Agenda.

The objective of the seminar was to encourage the participation of several researchers, teachers and professionals from national and international countries who conduct research on camels to present and debate, in plenary and conference sessions, their recent findings concerning this emblematic species, which is represented by the dromedary. This includes the varied and vital roles it has played in the past and continues to play today, its socio-economic and ecological implications, and the enhancement and improvement of products obtained from this species such as meat and milk. As a final point, experts have recommended the creation of a national camel research community of comprising academics, investors, promoters and professionals, with the establishment of a Maghreb network bringing together all actors in the camel research sector to form a common database. This will help to coordinate research projects in Algeria and throughout the Maghreb; the establishment of stations, research centers and camel breeding in potential regions; the introduction of an intensive system in parallel with the extensive breeding system to improve the sustainability of camel breeding; the encouragement of scientific studies, in particular in genetics, which make it possible to determine camel breed standards in Algeria and to adopt the same nomenclature throughout the national territory and worldwide. It will also contribute to the identification and characterization of different types of camels in relation to their abilities; the promotion of camel products such as meat, milk, skin and hair; the dietary and therapeutic virtues of camel products; studies on feed and nutritional needs according to the evolution of livestock systems; and a focus on dromedary pathologies, in particular parasitic pathologies and screening for zoonotic diseases. This book concludes that the preservation and the development of are a high priority to strengthen the exchange of information between experts of camel research sector. We wish to acknowledge all the members of the Organization Committee, the support of all the participants and sponsors who enabled the organization of the 10th International Seminar of Medicine: Camelids in Algeria and the Maghreb, and the Scientific Committee that ensures the scientific quality of these papers.

check for updates

Citation: Dib, A.L.; Boukhechem, S.; Bougherara, H.; Bererhi, E.-H. Preface of the 10th International Seminar of Veterinary Medicine: Camelids in Algeria and the Maghreb. *Biol. Life Sci. Forum* **2023**, 22, 2. https://doi.org/10.3390/blsf2023022002

Published: 30 March 2023

Conflicts of Interest: The authors declare no conflict of interest.

biology and life sciences forum

MDPI

Editorial

Statement of Peer Review

Amira Leila Dib *, Said Boukhechem, Hithem Bougherara and El-Hacene Bererhi

Gestion Santé et Productions Animales Research Laboratory, Institut des Sciences Vétérinaires El Khroub, Université Frères Mentouri Constantine 1, Constantine 25000, Algeria; said.boukhechem@umc.edu.dz (S.B.); heithem1985@yahoo.fr (H.B.); brerhihacene@yahoo.fr (E.-H.B.)
* Correspondence: dib.amiraleila@umc.edu.dz

In submitting conference proceedings to *Biology and Life Sciences Forum*, the volume editors of the proceedings certify to the publisher that all papers published in this volume have been subjected to their peer review. Reviews were conducted by expert referees to meet the professional and scientific standards expected of a proceedings journal.

- Type of peer review: single-blind;
- Conference submission management system: submission by email and Google docs
- Number of submissions sent for review: 38;
- Number of submissions accepted: 22 (selected based on the quality of presentation and scientific committee evaluation);
- Acceptance rate (number of submissions accepted/number of submissions received): 57%;
- Average number of reviews per paper: 2;
- Total number of reviewers involved: 36;
- Any additional information on the review process: no additional comments.

Conflicts of Interest: The authors declare no conflict of interest.

check for updates

Citation: Dib, A.L.; Boukhechem, S.; Bougherara, H.; Bererhi, E.-H. Statement of Peer Review. *Biol. Life Sci. Forum* **2023**, 22, 1. https://doi.org/10.3390/blsf2023022001

Published: 29 March 2023

biology and life sciences forum

MDPI

Proceeding Paper

The Camel in Algeria: Animal of the Past, Present and Future: What Is the Scope of Farming Systems? †

Abdelhakim Senoussi *, Aïcha Abazi, Saïd Bezziou and Zakaria Brahimi

Laboratoire de Recherche Bioressources Sahariennes: Préservation et Valorisation, Faculté des Sciences de la Nature et de la Vie, Université Kasdi Merbah Ouargla, Ouargla 30000, Algeria; abaziab2015@gmail.com (A.A.); bezziou.said69@gmail.com (S.B.); zakaria.brahimi096@gmail.com (Z.B.)

* Correspondence: senoussi.hakim@univ-ouargla.dz or senoussi.hakim@yahoo.fr

† Presented at the 10th International Seminar of Veterinary Medicine: Camelids in Algeria and Maghreb, Constantine, Algeria, 20–21 December 2022.

Abstract: In Algeria, the camel is one of the greatest riches and resource reservoirs of the Saharan territory. Compared to other farm animals, this species was relegated to the background, despite a past that testifies to a preponderant role in a hostile environment. Due to its legendary sobriety, it is the emblematic animal of the merchant caravan and is renowned for its versatility. It appears as an excellent means of locomotion, saddle, with interesting traction capacities in addition to being endowed with remarkable strength for plowing and the use of noria or delou. Today, the increase in numbers has been accompanied by changes in production systems, while the notable increase in the demand for camel products revealed by the socio-economic changes in nomadic communities, in terms of education and health, has contributed to modifying animal husbandry practices. Indeed, the emergence of new specialized camel systems has disrupted the habits that are now reflected in an increased interest in both the systems adopted and the products related to them. The future of the dromedary would be projected according to dimensions of renewal; the camel sector, local products, and camel services are many assets to consider. It is in this perspective that the present contribution, as a synthesis, attempts to highlight camel breeding in the Algerian Northern Sahara.

Keywords: Algeria; dromedary; livestock systems; vocation

Citation: Senoussi, A.; Abazi, A.; Bezziou, S.; Brahimi, Z. The Camel in Algeria: Animal of the Past, Present and Future: What Is the Scope of Farming Systems? *Biol. Life Sci. Forum* **2023**, 22, 3. https://doi.org/10.3390/blsf2023022003

Academic Editors: Amira Leila Dib, Said Boukhechem, Hithem Bougherara and El-Hacene Bererhi

Published: 30 March 2023

1. Introduction

The dromedary, without which the great nomadic civilizations could never have existed, occupies a predominant place in the economic and social life of the Saharan communities, which lived in harmony with their environment, yet were characterized by extremely vigorous living conditions. The caravan ensured the transport of goods as well as that of intrepid travelers and pilgrims, hoping to find in the "oasis islands" something to quench their thirst, eat their fill, and recover their strength [1]. This livestock breeding, long considered a survival of a bygone era, is of definite ecological, economic, social, and cultural interest.

What significance should be attributed to camel breeding in Algeria in the third millennium? Because of its qualities as an ambulatory walker, its legendary sobriety, and its habit of making do with the poor fodder of the desert, the dromedary is the emblematic animal par excellence because of its robustness, productivity, and diligence, it demonstrates all the qualities of a multi-purpose animal. The threshold of the 2000s constitutes a new boom for this animal marked by a new fate of the adopted systems approaching more urban areas via new marketing channels [2].

It is through this dimension that the present synthesis highlights the scope of camel breeding in the Northern Sahara. The diversity of its populations, the increase in its numbers, and the camel vocations are combined with new systems and give rise to the emergence of camel sectors of commerce. Faced with climate change and food security challenges, Algeria should count on this species as an important part of its national economy.

2. Materials and Methods

The method adopted highlights the camel system approach used in the Algerian Northern Sahara represented by regions of El-Oued, Ouargla, and Ghardaïa. Field surveys involving various stakeholders (camel drivers, producers, traders, and consumers) have made it possible to identify the trajectories and typology of livestock farming systems, their vocations, and the related sectors.

The objective of this study is to situate camel breeding practices.

3. Results and Discussion

Data mining and field investigations have enabled us to establish the following:

3.1. Size of Camels

Over a century, there was a sharp decline in population numbers, dropping from 260,000 heads in 1890 to 113,900 in 1988, while the decade that followed was marked by a jagged trajectory, reaching 154,310 heads in 1998 [3]. What can we learn from these fluctuations? The lack of interest in this species despite the boom in Saharan regions, but modernism deemed antinomic to dromedary [4].

At the dawn of the new millennium, a new impetus was given to camel breeding, with the number of camels increasing from 220,000 in 1999 to 435,214 in 2020 [3]. What can be deduced from this significant increase in the number of camels? The spatiotemporal anchoring of camel herds in and around urban centers as a result of the sedentarization of communities, the birth premium initiated through the National Fund for Agricultural Regulation and Development in 2000, and the increased demand for camel milk seem to be the main causes of the renewed interest in camel farming.

3.2. Specialized Breeding Systems

At present, in the Northern Sahara, where camel breeding is proving to be a booming business, reveals that livestock farms depend on a feeding system through which camel breeders adopt three distinct modes: grass-fed (Figure 1), stall-fed (Figure 2), or mixed. Each of the systems responds to specific functions, the analysis of which gave rise to four types of livestock farming according to their purpose: (i) dairy type, (ii) meat type, (iii) poly vocational type, and (iv) meharis for socio-cultural and sporting purposes.

Figure 1. Grazing Camel—Ghardaïa Region.

Figure 2. Tidjane Farm—El-Oued Region.

3.3. Multi-Purpose Camel

It is through the great string of its products and services that new prospects for the camel sector are opening up. A range of functions could not exist without the presence of this animal, which, thanks to its polyfunctionality, renders enormous services to camel drivers whose lives are intimately linked to the animal [5]. Consequently, an arrangement of camel breeding to better increase its production can be considered in the immediate future [6].

3.4. Camel Industry

Many products and by-products made from this animal can foster a real natural industry. Between meat and milk, camel hair is highly sought after for a range of textile products with strong cultural identities. Further, camel skin is also destined to develop a small industry [7]. Finally, the dung, characterized by a low nitrogen composition but rich in indigestible fibers, is suitable for processing into paper pulp [8].

3.5. Lucrative Uses

Useful for many functions, the camel is now part of the daily life of Saharan societies thanks to its sporting performances as a racing animal (Méhari), cultural, as a chessboard of festive circumstances (contests and games) and tourist, as a walking tool (Figure 3).

Figure 3. Racing Camel—Ouargla Region.

4. Conclusions

If sedentarization had made it possible to judiciously combine agricultural production and camel breeding by making the most of scarce resources (cultivable space and water points), then this association would be the culmination of a long process that has contributed to the intensification of its components. Validly, camel breeding systems are undergoing profound changes, moving from the grass system (divagation) to mixed systems (transhumant) to industrial forms (intensive), whose cameline sectors are beginning to organize themselves via promoting products with high economic value.

Author Contributions: Conceptualization, A.S.; methodology, A.S., A.A., S.B. and Z.B.; software, A.S., A.A., S.B. and Z.B.; validation, A.S., A.A., S.B. and Z.B.; formal analysis, A.S., A.A., S.B., and Z.B.; investigation, A.A., S.B. and Z.B.; data curation, A.S., A.A., S.B. and Z.B.; writing—original draft preparation, A.S.; writing—review and editing, A.S.; visualization, A.S.; supervision, A.S.; project administration, A.S. All authors have read and agreed to the published version of the manuscript.

Funding: This research received no external funding.

Institutional Review Board Statement: Not applicable.

Informed Consent Statement: Not applicable.

Data Availability Statement: The works consulted are detailed in the bibliography.

Conflicts of Interest: The authors declare no conflict of interest.

References

1. Marouf, N. *Lecture de l'espace Oasien*; FeniXX réédition numérique (Sindbad): Paris, France, 1980; p. 281. (In French)
2. Faye, B.; Senoussi, H.; Jaouad, M. Le dromadaire et l'oasis: Du caravansérail à l'élevage périurbain. *Cah. Agric.* **2017**, *26*, 14001. [CrossRef]

3. FAOSTAT. Food and Agriculture Data. Available online: http://www.fao.org/faostat/fr/#data/QL (accessed on 20 November 2022).
4. Faye, B.; Bengoumi, M.; Barkat, A. Le développement des systèmes camelins laitiers péri-urbains en Afrique. In Proceedings of the Atelier International sur le lait de Chamelle en Afrique, Niamey, Niger, 5–8 November 2003; FAO-CIRAD KARKARA: Rome, Italy, 2004; pp. 115–125. (In French)
5. Adamou, A. Note sur la polyfonctionnalité de l'élevage camelin. *J. Algér. Rég. Arid.* **2009**, *8*, 108–122. (In French)
6. Saadoud, M.; Nefnouf, F.; Hafaoui, F.Z. La viande cameline dans deux régions du Sud Algérien. *Viandes et Produits Carnés* **2019**. Available online: https://www.viandesetproduitscarnes.fr/index.php/fr/1020-la-viande-cameline-dans-deux-regions-du-sudalgerien (accessed on 15 November 2022).
7. Senoussi, A. Camel breeding in Algeria: Myth or reality? In Proceedings of the 19èmes Rencontres Autour des Recherches sur les Ruminants, Paris, France, 5–6 December 2012; I.N.R.A.: Paris, France, 2012; p. 308. (In French)
8. Senoussi, A. Le Camelin; Facteur de la Biodiversité et à Usages Multiples. In Proceedings of the Actes Séminaire International sur la Biodiversité Faunistique en Zones Arides et Semi Arides, Ouargla, Algérie, 20–22 October 2009; Université Kasdi Merbah: Ouargla, Algérie, 2009; pp. 265–273. (In French)

biology and life sciences forum

MDPI

Proceeding Paper

Place of the Dromedary in the Ecological Balance of Its Saharan Ecosystem †

Abdelmadid Chehma [1,*], Hassen Mahma [1], Hafida Trabelsi [1] and Johann Huguenin [2]

1 Laboratoire de Bio-Ressources Sahariennes: Préservation et valorisation, Université Kasdi Merbah Ouargla, Ourgla 30000, Algeria; hassen.mahma@yahoo.fr (H.M.); tr.hafida@yahoo.fr (H.T.)
2 UMR SELMET, CIRAD, F-34398 Montpellier, France; johann.huguenin@cirad.fr
* Correspondence: chehma.ab@univ-ouargla.dz
† Presented at the 10th International Seminar of Veterinary Medicine: Camelids in Algeria & Maghreb, Constantine, Algeria, 20–21 December 2022.

Abstract: The dromedary is the only breeding species capable of adapting to the harshness of its desert environment. It manages to reproduce, making the most of the very meagre floristic resources of the Saharan ecosystem. Moreover, thanks to these adaptive faculties and its particular feeding behaviour, it contributes largely to the preservation and proliferation of the floristic cover of its Saharan rangelands. Indeed, an overview of studies conducted by our research team has shown that this animal lives in perfect harmony with its environment. On the one hand, it adopts a mode of ambulatory grazing that respects the balance of floristic and vegetative diversity: (1) by being very selective with regard to the species and parts of the plant; (2) by moving from one plant to another without exhaustion and without uprooting; (3) by grazing the parts of the plant (stem leaves, seeds) according to availability; (4) by managing to cover its daily needs, whatever the variations in fodder supply (linked to seasonal climatic variations); and (5) by travelling daily distances that can exceed 50 km/day, so as not to exhaust and overload restricted spaces. On the other hand, the camel, by endozoochory, could play an important role in the long-distance seed dispersal of many wild plant species, and contribute to the maintenance of the diversity and propagation of its floristic cover. After gut passage, the germination percentage of seeds with physical dormancy increases and faeces represent a significant source of organic matter that is favorable for germination and seedling growth until favorable conditions (rainfall) for their germination occur.

Keywords: dromedary; Saharan ecosystem; preservation; flora; endozoochory

Citation: Chehma, A.; Mahma, H.; Trabelsi, H.; Huguenin, J. Place of the Dromedary in the Ecological Balance of Its Saharan Ecosystem. *Biol. Life Sci. Forum* **2023**, 22, 7. https://doi.org/10.3390/blsf2023022007

Academic Editors: Amira Leila Dib, Said Boukhechem, Hithem Bougherara and El-Hacene Bererhi

Published: 31 March 2023

1. Introduction

The dromedary lives in a very harsh and restrictive desert environment, where climatic conditions are at their most severe, characterized, in particular, by very high temperatures and very low and irregular rainfall [1,2]. This results in a well-adapted, very thin and very poor (very lignified) floristic cover [3]. This vegetation is the most perceptible witness of the climate [4]. The dromedary is the only livestock species capable of exploiting and enhancing these vast Saharan expanses. Several studies have highlighted the animal's anatomical and physiological adaptation to desertification, enabling it to conserve its energy [5], go without drinking for many weeks [6], recycle its nitrogen and make do with a poor diet. Moreover, because of its feeding behaviour, the dromedary grazes in such a way as to preserve its ecological environment [7]. It does not overgraze any type of vegetation and can reach the upper layers of plant formations, it does not strip the soil and the topsoil does not volatilize when trampled [8]. With a prospect of continuing this work, and in order to have data specific to camel farming in Algeria, our research team has set itself the objective of focusing on the relationship between the animal and its environment, by studying the contribution of the camel to the ecological balance of its Saharan ecosystem. This has been approached in two ways: the first is the daily monitoring of the camel in its pasture, in order to study the impact of its feeding behaviour (species and plants grazed, quantities, parts and organs of plants, grazing duration, distances travelled, etc.) on the preservation of the floristic cover of its Saharan ecosystem, and the second is to study the role of the dromedary in the seed dispersal and germination of wild Saharan plants by endozoochory (diversity of seeds collected from camel faeces, seed germination after passage through the digestive tract, in vitro and in vivo, etc.).

2. Materials and Methods

To monitor the feeding behaviour of the dromedary, we directly observed it during grazing, based on video sequences recorded during the day (morning and afternoon). The diet was determined by the bite method, inspired by the work of [9,10].

For seed dispersal by endozoochory, we collected camel faeces from the dromedary's rangeland and analysed their seed contents, which were then subjected to germination tests (in vitro and in vivo) and compared to seeds directly taken from the parent species in the field.

3. Results and Discussion

Our study on the feeding behaviour of the dromedary showed that its activities on rangelands were directly affected by the availability of fodder, related to seasonal variation. In general, the dromedary grazed 28 species, amounting to 82% of the species present on the rangeland. However, the diet was more diversified in the autumn and spring (12 and 19 species) than in the winter and summer (five and four species). It devoted 67% of its time to grazing in the cooler seasons, compared to only 34% in the warmer seasons. The most abundant species were generally the most grazed (p-value = 0.001). It grazed in groups, especially in relation to trees or large plants. It could travel up to 50 km/d and constantly wandered around the foot of plants, taking small bites from each individual one. The duration of grazing, per plant, depended on its architecture and size (p-value of 0.000), thus favouring the revitalization of its space. In spite of the spatio-temporal variations in the fodder supply of its rangelands, it composed rations that covered its nutritional needs. The daily quantity grazed varied from 0.42 to 3.71 kg of dry matter/100 kg live weight/day, depending on the season and the floristic procession of the rangeland (p-value 0.00). An analysis of the rations consumed showed that the dromedary, thanks to its selective aptitudes, was capable of self-sustaining its nutritional needs (UFL and PDI) whatever the vegetation present and whatever the season and the rangeland. To that end, seasonal and spatial (rangeland) variations were not significant for the nutrient supply of the rations consumed, recording respective p-values of 0.684; 0.202 for energy, and 0.623; 0.128 for nitrogen.

Regarding the role of the camel in seed dissemination, we collected 39 species from faeces (in good condition), belonging to 18 botanical families. At the same time, 22 species belonging to nine botanical families retained their germination power after passage through the digestive tract. In addition, faeces offered a favourable microclimate and amounted to a significant source of organic matter favouring the growth of seedlings. Finally, it can be deduced that the dromedary, because of its ambulatory feeding behaviour over very long distances, remains the best animal species for dispersing seeds in the Saharan environment.

4. Conclusions

The data from our study showed that the dromedary, because of its particular feeding behaviour, contributes largely to the maintenance and proliferation of its floristic surroundings by a reasoned selection and grazing of species according to their availability and abundance, and the dissemination, lifting of dormancy and dispersion over long distances of the seeds of wild Saharan plants, etc. This led us to conclude that it is an animal that lives in harmony with its natural environment, and that it is the only livestock species capable of exploiting and enhancing vast desert spaces.

Author Contributions: Conceptualization, A.C. and J.H.; methodology, A.C., H.M., H.T. and J.H.; software, H.M. and H.T.; validation, A.C., H.M., H.T. and J.H.; formal analysis, A.C., H.M., H.T. and J.H.; writing—review and editing, A.C., H.M., H.T. and J.H.; supervision, A.C. All authors have read and agreed to the published version of the manuscript.

Funding: This research received no external funding.

Institutional Review Board Statement: The study did not require ethical approval.

Informed Consent Statement: Not applicable.

Data Availability Statement: Not applicable.

Conflicts of Interest: The authors declare no conflict of interest.

References

1. Seltzer, P. *Le Climat de l'Algérie*; Imprimerie "La Typo-litho" & J. Carbonel: Algiers, Algeria, 1946. (In French)
2. Ozenda, P. *Flora and vegetation of the Sahara*; CNRS Éditions: Paris, France, 1991.
3. Chehma, A.; Youcef, F. Seasonal variations of the floristic characteristics and chemical composition of Saharan routes in Southern Algeria. *Sci. Changements Planétaires/Sécher.* **2009**, *20*, 373–381. [CrossRef]
4. Wilson, R. *The Camel*; Longman: London, UK, 1984; 223p.
5. Bengoumi, M.; Faye, B. Adaptation du dromadaire à la déshydratation. *Sci. Changements Planétaires/Sécher.* **2002**, *13*, 121–129. (In French)
6. Chehma, A.; Longo-Hammouda, F.-H. Bilan azoté et gain de poids, chez le dromadaire et le mouton, alimentés à base de sous-produits du palmier dattier, de la paille d'orge et du drinn "Aristida pungens". *Cah. Agric.* **2004**, *13*, 221–226. (In French)
7. Newman, D. The feeds and feeding habits of old and new world camels. In *The camelid. An all-purpose animal. Volume I. Proceedings of the Khartoum workshop on camels, December 1979*; Cockrill, W.R., Ed.; Nordiska Afrikainst: Uppsala, Sweden, 1984; p. 250. (In French)
8. Stiles, D. Le dromadaire contre l'avancée du désert. *Recherche* **1988**, *201*, 948–952. (In French)
9. Meuret, M. Une histoire de coups de dents. *Rev. Spéc. Deséleveurs Chèvres* **1985**, *151*, 21–27. (In French)
10. Kassilly, F.N. Forage quality and camel feeding patterns in Central Baringo, Kenya. *Livest. Prod. Sci.* **2002**, *78*, 175–182. [CrossRef]

biology and life sciences forum

MDPI

Proceeding Paper

An Overview of Camel Biodiversity and Genetics †

Semir Bechir Suheil Gaouar [1,*] and Elena Ciani [2]

1 Applied Genetic in Agriculture, Ecology and Public Health (GenApAgiE) Laboratory, Faculty SNV/STU, University of Tlemcen, Tlemcen 13000, Algeria
2 Department of Biosciences, Biotechnologies and Biopharmaceutics, University of Bari "Aldo Moro", 70121 Bari, Italy
* Correspondence: suheilgaouar@gmail.com
† Presented at the 10th International Seminar of Veterinary Medicine: Camelids in Algeria & Maghreb, Constantine, Algeria, 20–21 December 2022.

Abstract: The process of desertification affects about 46% of Africa. Hence, the dromedary appears by far to be the most appropriate strategic investment. It is obvious that in view of global events, the number of areas that will be affected by desertification will increase considerably, which will make this animal a key element in maintaining food security for a considerable number of people. The objective of this review is to describe what is currently being carried out in the study of the phenotypic and molecular diversity of camels and to highlight the prospects for the exploitation and use of these products in sustainable farming systems. Due to the increasing demand for camels around the world, there is a need for knowledge about their phenotypic and genetic diversity. This is fundamental to the sustainable management and use of herds. It appears through this review that the morphological and genetic diversity of African camels and the scientific advance can be used for genetic improvement and conservation.

Keywords: Africa; camel; diversity; morphology; genetic

check for updates

Citation: Gaouar, S.B.S.; Ciani, E. An Overview of Camel Biodiversity and Genetics. *Biol. Life Sci. Forum* **2023**, 22, 15. https://doi.org/10.3390/blsf2023022015

Academic Editors: Amira Leila Dib, Said Boukhechem, Hithem Bougherara and El-Hacene Bererhi

Published: 10 April 2023

1. Introduction

Improving the diversification of locally adapted species and improving the ability to use genotypes to accurately predict relevant adaptive and production phenotypes has become a necessity with climate change [1,2]. As a result, camels appear to be the key animal in the situation. Camels are renowned for their production of quality milk, meat and fiber [3,4], are key species for ecological sustainability [5], and are sustainable with specific attributes, including draught capacity, means of transport and ecotourism, thus contributing to economic empowerment and food security. Understanding the structure and function of genomes is important for studying interactions between genes, studying interactions between genes and environments, and deciphering trait complexes. A high-quality reference genome assembly is a prerequisite to initiating functional genome annotation. Significant progress has been made in this direction by sequencing whole animal genomes, detecting sequence variants, associating them with phenotypic traits, and using genomic variation to select for predicted genetic differences in routinely measured traits [6]. Since characterization of a breed is the first step in its sustainable use; the extent of phenotypic and genetic variation is fundamental to the selection, improvement and use of diverse camel populations in Africa [7].This review covers several characteristics of camels in Africa.

2. Morphological Diversity

The distinction between camel populations by herders is subject above all to tribal detention. However, different intrapopulation classifications exist (notably through the color of the coat and the vocation of the animal). This parameter, in addition to being a selection criterion for breeders, also plays an important role in terms of zootechnical performance [8]. The selection of animals according to their colors is rather influenced by culture and not zootechnical yield; for the latter, the selection of breeders is based on other criteria [3]. However, Amine et al. [3] provides evidence that skin color has a perceptible influence on the zootechnical performance of camels. This was confirmed genetically by the study by Holl et al. [8]. It appears from these studies and many others that the genes that control this phenotype have a significant pleiotropic effect that affects zootechnical parameters.

3. Molecular Diversity

Almathen et al. [9] conducted a study around the world with microsatellites and mtDNA markers and found no clear phylogenetic patterns. Apparently, the movements of camels along transcontinental trade routes may have eroded pre-existing phylogeographic patterns, resulting from their initial domestication. The same finding was highlighted by Cherifi et al. [10] with microsatellites on Algerian and Egyptian animals. The recent domestication of the dromedary, and selective pressures are factors that can be at the origin of a such result. However, with the development of the economic activities around this species, populations have started to be distinguished in terms of dairy, butchering and sports skills.

4. Genetic Studies

Building on newly available high-quality reference genomes, polymorphism analyses, both structural and functional, and the first genome-wide association studies, many outcomes were possible. These include the identification of genomic regions of environmental adaptation and heat stress, and the identification of genes and quantitative trait loci (QTL).

5. Conclusions

In terms of biodiversity; the camel species remains a species that contains significant genetic variability, especially at the intrapopulation level. This implies a great predisposition to selection. In terms of genetics, significant progress has been made in the knowledge of the multiple facets of dromedary genomics, which implies a better understanding of the improvement of production in this species.

Author Contributions: Conceptualization, S.B.S.G.; methodology, investigation, S.B.S.G.; writing—original draft preparation, S.B.S.G.; writing—review and editing, S.B.S.G.; visualization, S.B.S.G.; supervision, S.B.S.G. and E.C. All authors have read and agreed to the published version of the manuscript.

Funding: This research received no external funding.

Institutional Review Board Statement: Not applicable.

Informed Consent Statement: Not applicable.

Data Availability Statement: The data presented in this study are available upon request from the corresponding author.

Conflicts of Interest: The authors declare no conflict of interest.

References

1. Reich, P.F.; Numbem, S.T.; Almaraz, R.A.; Eswaran, H. Land resource stresses and desertification in Africa. In *Response to Land Degradation*; CRC Press: Boca Raton, FL, USA, 2001; pp. 101–116.
2. Clark, E.L.; Archibald, A.L.; Daetwyler, H.D.; Groenen, M.A.M.; Harrison, P.W.; Houston, R.D.; Kühn, C.; Lien, S.; Macqueen, D.J.; Reecy, J.M.; et al. From FAANG to fork: Application of highly annotated genomes to improve farmed animal production. *Genome Biol.* **2020**, *21*, 285. [CrossRef] [PubMed]
3. Amine, C.Y.; Samir, G.S.B.; Nasreddine, M.; Nacera, T.A.; Nadhira, S.M. Study of camelina biodiversity in southwestern of Algeria. *J. Life Sci.* **2013**, *7*, 416–427.

4. Dich, I.; Abdelbari, H.; Fardeheb, L.; Dich, S.; Gaouar, S.B.S. Characterization of the fineness of she-camel's wool in the Wilaya of Nâama and El Bayadh. *Genet. Biodivers. J.* **2022**, *6*, 181–195. [CrossRef]
5. Slimani, N. Impact du comportement alimentaire du dromadaire sur la préservation des parcours du Sahara septentrional algérien: Cas de la région de Ouargla et Ghardaïa. PhD Thesis, Université Kasdi Merbah, Ouargla, Algeria, 2015. (In French).
6. A Burger, P.; Ciani, E. Structural and functional genomics in Old World camels—where do we stand and where to go. *Anim. Front.* **2022**, *12*, 30–34. [CrossRef] [PubMed]
7. Gaouar, S. Contribution à l'étude de la variabilité génétique des races ovines par l'utilisation des microsatellites: Caractérisation de deux races ovine algériennes Hamra et Ouled-Djellal. Master Thesis, Laboratoire de Biologie Moléculaire et Génétique, Institut des Sciences de la Nature, Université d'Es-Sénia, Oran, Algeria, 2002. (In French).
8. Holl, H.; Isaza, R.; Mohamoud, Y.; Ahmed, A.; Almathen, F.; Youcef, C.; Gaouar, S.; Antczak, D.F.; Brooks, S. A Frameshift Mutation in KIT is Associated with White Spotting in the Arabian Camel. *Genes* **2017**, *8*, 102. [CrossRef] [PubMed]
9. Almathen, F.; Charruau, P.; Mohandesan, E.; Mwacharo, J.M.; Orozco-terWengel, P.; Pitt, D.; Abdussamad, A.M.; Uerpmann, M.; Uerpmann, H.-P.; De Cupere, B.; et al. Ancient and modern DNA reveal dynamics of domestication and cross-continental dispersal of the dromedary. *Proc. Nat. Acad. Sci. USA* **2016**, *113*, 6707–6712. [CrossRef] [PubMed]
10. Cherifi, Y.A.; Gaouar, S.B.S.; Guastamacchia, R.; El-Bahrawy, K.A.; Abushady, A.M.A.; Sharaf, A.A.; Harek, D.; Lacalandra, G.M.; Saïdi-Mehtar, N.; Ciani, E. Weak Genetic Structure in Northern African Dromedary Camels Reflects Their Unique Evolutionary History. *PLoS ONE* **2017**, *12*, e0168672. [CrossRef] [PubMed]

biology and life sciences forum

MDPI

Proceeding Paper

Comparative Study of the Phenotypic and Molecular Genetic Diversity of "Tergui" Camel Population in the Hoggar Region (South Algeria) †

Derradji Harek [1,2,*], Hacene Ikhlef [2], Youcef Amine Cherifi [3], Rachid Bouhadad [4] and Fodil Arbouche [5]

1 Higher National Agronomic School-ENSA (ex. INA), Algiers 16000, Algeria
2 National Institute of Agronomic Research of Algeria (INRAA), Algiers 16010, Algeria; h.yakhlef@ensa.dz
3 Molecular and Cellular Genetics Laboratory, University of Science and Technology of Oran Mohamed-Boudiaf USTOMB, Oran 31000, Algeria; cherifi.youcef@ymail.com
4 Laboratoire Ecologie Animale, Université des Sciences et de la Technologie Houari Boumediene USTHB, Bab Ezzaouar, Algiers 16111, Algeria; rbouhadad@yahoo.fr
5 Department of Agronomy, University of Ghardaia, Ghardaia 47000, Algeria; arbouchefodil@yahoo.fr
* Correspondence: derradji11@gmail.com
† Presented at the 10th International Seminar of Veterinary Medicine: Camelids in Algeria & Maghreb, Constantine, Algeria, 20–21 December 2022.

Abstract: In the framework of the sustainable management of the genetic diversity (morphometric and molecular) of the "Tergui" camel population represented by the phenotypes (Abahou, Amelal, Alemlagh, Atelagh, and Azerghaf), surveys in the Hoggar area of 87 individuals from 11 localities were conducted in order to estimate the variability of the inter-intra-population. The morphological results provide full information about the structure of this population and demonstrate an important polymorphism. The results of the genotyping of the DNA with 20 microsatellite markers made it possible to demonstrate inter- and intra-population genetic variability characterized by a high rate of heterozygosity (Hnb) and effective alleles. The rate of heterozygosity in our phenotypes varied from 0.56 to 0.63, which is higher than that observed in foreign populations ranging from 0.537 to 0.629. A total of 169 alleles of 20 microsatellite loci were detected. The mean number of alleles per locus was 7.15, 6.15, 3.10, 4.45, and 3.25 for Abahou, Amelal, Alemlagh, Atelagh, and Azerghaf, respectively. The loci evaluation showed higher PIC values greater than 0.5, which are considered very instructive. The heterozygous values observed for all the loci analyzed were lower than expected, which could be attributed to inbreeding in the population or subdivision of the studied population into distinct breeds and phenotypes. On the other hand, the number of observed alleles is higher and has shown a frequency that exceeds 7.3%. The genetic differentiation values between the phenotypes analyzed were much lower and the level of differences accounted for 1.1% of the total genetic variation. All loci contributed to this differentiation with FST values being moderately low and similar but very significant ($p < 0.001$). The overall FST value was similar but slightly higher than that of 0.9%. The genetic similarity between the phenotypes and the classification methods (AFC and DACP) gave results similar to the phenotypic characteristics, and showed that they appear to be genetically very similar, thereby supporting the decision to consider them only mildly differentiated.

Keywords: dromedary; genetic characterization; genetic variability; microsatellite; *Camelus dromedarius*; Tamanrasset

check for updates

Citation: Harek, D.; Ikhlef, H.; Cherifi, Y.A.; Bouhadad, R.; Arbouche, F. Comparative Study of the Phenotypic and Molecular Genetic Diversity of "Tergui" Camel Population in the Hoggar Region (South Algeria). *Biol. Life Sci. Forum* **2023**, 22, 6. https://doi.org/10.3390/blsf2023022006

Academic Editors: Amira Leila Dib, Said Boukhechem, Hithem Bougherara and El-Hacene Bererhi

Published: 30 March 2023

1. Introduction

Genetic resources are the most valuable and strategically important commodity because many indigenous populations, races, and ecotypes contribute to human needs and could generate much more than they currently do. In order to ensure the sustainability of species and avoid their extinction, the study and management of biodiversity allows us to accumulate knowledge and orient towards their economic value on the genetic level and to qualify biodiversity as a potential source of permanent income on a global scale [1].

An effort has been made to determine, through the analysis of genetic diversity, the different phenotypes (ecotypes) of the "Tergui" population that have until now only been differentiated based on phenotypic appearance, geographical location, or tribal property. It was agreed that variations in the main morphological criteria, genotype, and performance that make up populations were not sufficient to describe them as "races". This study deals with the camel's genetic variability; its main objective is to characterize the genetic diversity of the "Tergui" camel population in the Hoggar region based on the phenotypic and molecular criteria, to analyze it within and between ecotypes, and to highlight the prospects of its exploitation in systems of sustainable livestock.

2. Material and Methods

The study was carried out throughout the Hoggar region. Initially, a pre-survey in order to select the localities, range, and degree of diversity of the phenotypes encountered with maximum representativeness was conducted [2], covering 11 localities and a part of Tamanrasset: Tamanrasset, Izernene, Asskrem, Tin-Amzi, Tarahnent, Izerzi, In-Amguel, Hirafok, Silet, and Abalessa. The morphologic methodology was based on dimensionless dimensional space, and then sampling the DNA extraction of the blood samples of 87 animals.

3. Results and Discussion

The usefulness of microsatellite and phenotypic observed in all the ecotypes (Abahou, Amellal, Alemlagh, Atelagh, and Azzerghaf) had high heterozygosity values (0.62, 0.63, 0.62, 0.59, and 0.62, respectively). The number of alleles observed is higher, and most of private alleles (55) were at very low frequencies (below 3%), although two alleles unique to Abahou population CVRL7 (292) and CMS9 (237) showed a frequency that exceeded 7.3%. Genetic differentiation values among the five analyzed camel populations are much lower, and the level of differences explained 1.1% of the total genetic variation. All loci contribute to this differentiation, with F_{ST} values being moderately low and similar for all of the studied systems, and yet still very significant ($p < 0.001$). Our overall F_{ST} value was similar to but slightly higher than the 0.9%. However, it was smaller than those previously found between Indian camel breeds (F_{ST} = 8.2%, [3] and Tunisian camel populations (F_{ST} = 9%), [4]. The genetic similarity between the Algerian Tergui populations was further illustrated using genetic distances, correspondence analysis, and clustering methods (DAPC). These three classical estimates are based on genetic relationships, and gave similar results. The neighbor-joining tree (DR) showed a clear subdivision of the breeds into tree main groups, each divided into subgroups, and the Azerghaf population was more closely related to Abahou–Amelal group than the Alemlagh–Atelagh group. Our estimates for the relationship of the indigenous population is similar to the one reported by the phenotypic, where it has been pointed out using characteristics (color). Furthermore, the genetic similarity of the Tergui populations was also demonstrated using the FCA and the DAPC clustering approaches. These results could be explained by the two populations being common ancestors, with an extensive gene flow between the five Tergui populations.

4. Conclusions

This study on the genetic diversity of the camel's "Tergui" through the five phenotypes revealed great phenotypic diversity, which is due to the presence of a number of mutations with a visible effect. The genetic polymorphism was studied morphologically and molecularly. It appears the population that is geographically neighboring is genetically close, and there seems to be a significant divergence between geographically distant populations (cases of the Abahou and Azerghaf phenotypes). Moreover, the analysis as well as the genetic distances show that they appear to have a genetic kinship. The use of microsatellite markers made it possible to confirm the genetic diversity observed at the morphological level. The genotyping results presented, even though the number of loci studied is quite large, and provided an initial estimate of camel gene diversity. In addition, it shares a large genetic base, despite the appearance of regional differences with a fairly high inbreeding rate of 15%. Indeed, the population seems threatened by the lack of management of genealogies in the herds and by crossbreeding, which could lead to a change in its genetic structure towards homozygosity. It appears, then, that the lack of a livestock management program for a given type of production and the importance of the uncontrolled migration of individuals among the herds both cause a continuous flow of genes between the phenotypes.

Author Contributions: Conceptualization, D.H. and F.A.; methodology, D.H.; Y.A.C. and F.A. software, D.H.; validation, H.I.; F.A.; R.B. and D.H.; formal analysis, D.H.; investigation, D.H., F.A.; resources, D.H.; data curation, D.H.; writing—original draft preparation, D.H.; writing—review and editing, D.H.; visualization, D.H.; supervision, D.H.; project administration, D.H.; funding acquisition, D.H. All authors have read and agreed to the published version of the manuscript.

Funding: This research received no external funding.

Institutional Review Board Statement: Not applicable.

Informed Consent Statement: Not applicable.

Data Availability Statement: The data presented in this study are available upon request from the corresponding author.

Acknowledgments: This work was carried out within the framework of the surveys conducted in Algeria for the development of camel breeding, which was supported by the National Institute of Agronomic Research of Algeria (INRAA). We acknowledge the help of many colleagues from other institutions and universities. The authors are also grateful to the Direction des Services Agricole and the Chamber of Agriculture of the Tamanrasset Wilayate, which facilitated the logistics for the realization of the surveys, as well as to the camel breeders of the region who gave their time for the collection of data.

Conflicts of Interest: The authors declare no conflict of interest.

References

1. Food and Agriculture Organization of the United Nations. *Developing the Institutional Framework for the Management of Animal Genetic Resources*; FAO Animal Production and Health Guidelines, No.6; FAO: Rome, Italy, 2011.
2. Woolliams, G.A.; Pilling, D.; Scherf, B. Surveying animal genetic resources to manage biodiversity. *Anim. Genet. Resour.* **2010**, *47*, 23–30. [CrossRef]
3. Vijh, R.K.; Tantia, M.S.; Mishra, B.; Kumar, S.T.B. Genetic Diversity and Differentiation of Dromedarian Camel of India. *Anim. Biotechnol.* **2007**, *18*, 81–90. [CrossRef] [PubMed]
4. Ahmed, M.O.; Salem, F.B.; Bedhiaf, S.; M'Naouer, D. Analyse moléculaire de la diversité génétique des dromadaires (Camelus dromedarius) en Tunisie. *Biotechnol. Agron. Soc. Environ.* **2010**, *14*, 399–408.

biology and life sciences forum

MDPI

Proceeding Paper

Camel Breeding in Algeria †

Nassim Moula [1,2]

[1] Department of Animal Production, Fundamental and Applied Research for Animal and Health, Faculty of Veterinary Medicine, University of Liege, B-4000 Liege, Belgium; Nassim.Moula@uliege.be
[2] GIGA Animal Facilities, University of Liege, B-4000 Liege, Belgium
† Presented at the 10th International Seminar of Veterinary Medicine: Camelids in Algeria & Maghreb, Constantine, Algeria, 20–21 December 2022.

Abstract: Camel breeding remains marginal in Algeria but represents an invaluable animal resource for desert regions. The camel is one of the few domestic species that provides milk, meat, leather, wool, manure and work. The Algerian camel herd has experienced a significant increase in recent years, reaching a population of about 416,500 in 2019. This livestock is located in 17 Wilayates, including 8 Saharan and 9 Stepppiques. The main breeds raised in Algeria are: the Chaambi, Ouled Sidi Cheikh, Ait Khebbach, Steppe Camel, Saharaoui, Targui, Ajjer, Reguibi and Ftouh. The camel sector in Algeria is facing several constraints of a sanitary, genetic, logistical and organizational nature. The management of land and common spaces, following the sedentarization of some breeders, is another difficulty that the sheep sector must face. Several favorable factors for camel breeding in Algeria, such as the pedoclimatic diversity of the country, culture/religion, economic and genetic diversity, can help to improve Algerian red meat production. The objective of this paper is to present the current situation of camel farming in Algeria.

Keywords: dromedary; Algeria; livestock system; arid regions; Sahara

Citation: Moula, N. Camel Breeding in Algeria. *Biol. Life Sci. Forum* **2023**, 22, 4. https://doi.org/10.3390/blsf2023022004

Academic Editors: Amira Leila Dib, Said Boukhechem, Hithem Bougherara and El-Hacene Bererhi

Published: 30 March 2023

1. Introduction

Algeria is the largest African country, with an area of 2,381,741 km^2. It has great diversity in its pedoclimatic conditions and plant and animal genetic resources. Algeria is home to a diverse range of domestically produced animals, including chickens, camels, goats, cattle and sheep [1].

Arid and desert steppe ecosystems are marked by great landscape diversity in relation to significant variability in ecological factors. Regions with a pastoral tradition are populated with people who are pastoralist–breeders; most of them are former nomads, with a strong tendency for sedentarization today. The dromedary is a multifunctional animal with an excellent capacity to adapt to the conditions of life in arid and semi-arid environments. It can remain alive for a long period without drinking. It is also able to convert meager forage resources, usually untapped by other domestic animals, into energy [2]. Despite its economic, social and cultural importance, the Algerian camel has been neglected in development programs, and few studies are available on camel husbandry systems in the country.

This study aims to present the most up-to-date situation on camels in Algeria, based on official reports and scientific publications.

2. Current Situation of Camel Breeding in Algeria

The camel population in Algeria was estimated to be about 416,500 in 2019. As shown in Figure 1, the camel herd has experienced an average annual increase of about 4%. This number places Algeria in 14th place among countries breeding camels. In 1890, the number of camels in Algeria was estimated at 260,000. They decreased to 194,000 in 1910 and to 141,000 in 1986 [3]. The dromedary is present in 17 Wilayates (8 Saharan and 9 Steppic).

There are three main distribution areas. With 56% of the national livestock, the central Sahara is the major distribution area of the dromedary in Algeria. The two other areas are the Northern Sahara (37%) and the steppe (7%). The main camel breeds in Algeria are the Chaambi, Ouled Sidi Cheikh, Ait Khebbach, Chameau de la steppe, Saharaoui, Targui, Ajjer, Reguibi and Ftouh [3].

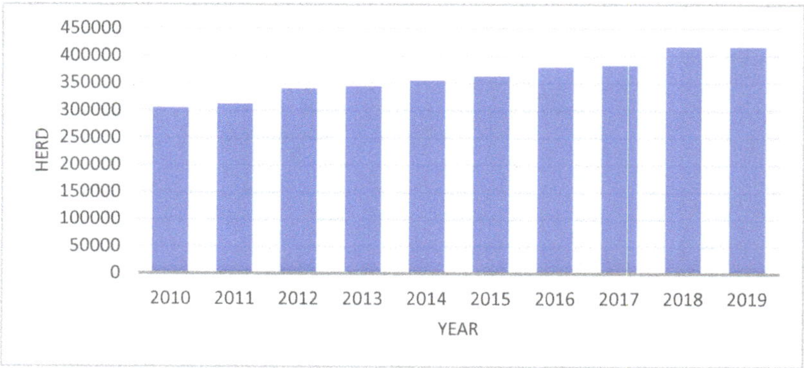

Figure 1. Evolution of the camel population in Algeria [4].

3. Dromedary Breeding Systems in Algeria

Traditionally, in camel farming, spatial mobility of the herds is the first strategy recommended for the search of sufficient feed and water for herds. The exploitation of pastures constitutes the basis of dromedary feeding. The feeding behavior of the dromedary contributes to a rational use of natural resources with greater variability in the consumed plants. The dromedary contributes significantly towards the fight against desertification, consuming a wide variety of forage species, both botanically and chemically, some of which are rejected by other domestic ruminants. The dromedaries are raised in Algeria following three systems of breeding: sedentary, nomadic and transhumant. However, due to the alarming change in the security situation in recent years, dromedary farms are mainly of a "sedentary" type with a strong exploitation of the pastures around the farms.

4. Use of Dromedary in Algeria

In general, breeders rely on four main qualities: (i) species diversity, (ii) adaptation to the environment, (iii) hardiness and (iv) versatility to constitute their herd. As in other countries, the use of dromedary in Algeria is very numerous. Milk is often consumed fresh and/or fermented by breeders. Some breeders mix the milk from their camels with that of goat milk to produce butter and fresh cheese, for example, Oudi ouanamellen, tikamarine (cheese) and Oudi ouanamellen (butter) among the Tuareg. These products are intended for self-consumption. Camel meat represents 3% of Algerian red meat production. The camel represents one-third of the slaughtered animals in the southern wilayas. The meat is often consumed boiled. Nomads are very fond of the hump fat. The oubar is a raw material that is required for manufacturing a whole range of products necessary to the nomadic life (bags, blankets, shackles). The dromedary hair has a very good market value. In Algeria, the burnous in oubar is very appreciated and expensive, especially when it is made with the hairs of young dromedary. The skin is usually thrown away. However, some Tuareg craftsmen take advantage of this opportunity to transform it into "iretmen", and good-quality soles are offered for sale in the city market at good prices. Other camel breeders, especially those of the Hoggar, continue to exploit it to obtain a whole range of products necessary for their daily life: "ahloum", ropes used to draw water, "ikchir", a kind of bag to keep certain supplies, as well as carpets. The dromedary is also used in Algeria for work, including plowing, transporting goods and people and drawing water from deep wells.

5. Conclusions

Camel breeding in Algeria can be improved through efforts at three levels: technical, commercial and institutional. At the technical level, improvements in farm productivity imply interventions in various fields, such as animal health, livestock feeding or even the technical management of herds. At the commercial level, it is essential to enhance the commercial value by improving the conditions required for better access to the markets. At the institutional level, an articulation of the interventions of the institutional actors involved in the field of camel breeding in Algeria must be improved to find sustainable solutions to the problems of the breeders.

Funding: This research received no external funding.

Institutional Review Board Statement: Not applicable.

Informed Consent Statement: Not applicable.

Data Availability Statement: Not applicable.

Conflicts of Interest: The authors declare no conflict of interest.

References

1. Moula, N. Caractérisation de la race ovine algérienne Tazegzawth. *Tropicultura* **2018**, *36*, 43–53. [CrossRef]
2. Faye, B.; Abdelhadi, O.; Raiymbek, G.; Kadim, I.; Hocquette, J.F. La production de viande de chameau: état des connaissances, situation actuelle et perspectives. *INRA Prod. Anim.* **2013**, *26*, 289–300. [CrossRef]
3. Aissa, B. Le dromadaire en Algérie. In *Séminaire sur la Digestion, la Nutrition et L'alimentation du Dromadaire*; Tisserand, J.-L., Ed.; Options Méditerranéennes: Series A. Mediterranean Seminars, OMA02; CIHEAM: Zaragoza, Spain, 1989; pp. 19–28.
4. MADR. *Ministère de L'agriculture et du Développement Rural: Statistiques Agricoles et Production Animale*; Ministère de l'Agriculture et du Développement Rural: Abidjan, Ivory Coast, 2019.

biology and life sciences forum

MDPI

Proceeding Paper

Breeding and Nutrition of Camels in Algeria †

Said Boukhechem [1], Amira Leila Dib [1,*], Hithem Bougherara [1], Assia Bouaziz [1], Louiza Kadja [1], Khireddine Ghougal [1], Ibtissem Oubira [1], Zahia Chari [1], Nessrine Sana Kerrour [2], Melisa Lamri [3], Nedjoua Lakhdara [1] and El-Hacene Bererhi [1]

1 Gestion Santé et Productions Animales Research Laboratory, Institut des Sciences Vétérinaires El- Khroub, Université Frères Mentouri Constantine 1, Constantine 25000, Algeria; said.boukhechem@umc.edu.dz (S.B.); heithem1985@yahoo.fr (H.B.); bouaziz.assialp@gmail.com (A.B.); louiza20132014@gmail.com (L.K.); epidemiologie25@gmail.com (K.G.); ibtissem.oubira@yahoo.fr (I.O.); manel_kaouther25@yahoo.fr (Z.C.); nedjoua2002@hotmail.com (N.L.); brerhihacene@yahoo.fr (E.-H.B.)

2 Department of Food Science, Université Abderrahmane Mira, Béjaia 06000, Algeria; sanakerrour@outlook.fr

3 Department of Food Science, Mouloud Mammery University, Tizi-Ouzou 15000, Algeria; lamrimeliza1@gmail.com

* Correspondence: dib.amiraleila@umc.edu.dz

† Presented at the 10th International Seminar of Veterinary Medicine: Camelids in Algeria & Maghreb, Constantine, Algeria, 20–21 December 2022.

Abstract: Among the mammals domesticated by humans for their needs, the dromedary, which is a versatile animal, is used for the production of milk, meat, wool and skin, as well as for sports, tourism and transport. However, knowledge about the breeding and feeding of this animal is limited. The objective of this work is to synthesize previous works on the breeding and feeding of camels in Algeria. The results show that the main farming systems are transhumant, sedentary and nomadic. In addition, the diet of the dromedary is based on different types of Saharan routes and constitutes the main element upon which an extensive camel breeding system is based. Thus, for a better valorization of this species, further studies on the diet and breeding methods of camels are recommended.

Keywords: camelids; Algeria; breed; nutrition; Sahara

check for updates

Citation: Boukhechem, S.; Dib, A.L.; Bougherara, H.; Bouaziz, A.; Kadja, L.; Ghougal, K.; Oubira, I.; Chari, Z.; Kerrour, N.S.; Lamri, M.; et al. Breeding and Nutrition of Camels in Algeria. *Biol. Life Sci. Forum* **2023**, *22*, 5. https://doi.org/10.3390/blsf2023022005

Academic Editor: Boussena Sabrina

Published: 30 March 2023

1. Introduction

The Camelidae is included in the genera *Camelus* (dromedary and Bactrian species) and includes the Llama (guanaco and llama) and the Vicuna (vicuna and alpaca). However, camel is often used to describe all the aforementioned camel-like animals. Furthermore, the one-humped dromedary (*Camelus dromedarius*) accounts for more than 90% of all camels, while the two-humped Bactrian camel (*Camelus bactrianus*) accounts for only 10% [1]. According to the FAO 2020, the total number of camels recorded in the world was 35,525,270 in 2018 and could reach 60 million in 25 years [2]. Additionally, Alaskar et al. [3] reported the following distribution of camels in the world: 33.02% in East Africa, 20.9% in Central Africa, 15.96% in North Africa, 5.27% in West Asia, 4.05% in Australia, 2.11% in East Asia, 0.94% in Central Asia, and 0.02% in Europe. Five neighboring countries, including Somalia, Ethiopia, Kenya, Sudan and Djibouti, hold 84% of Africa's camel herd and more than half of the world's camel herd. In addition, camel breeding system is different from one country to another, and their feed needs are not determined, with only a few recommendations being available. In addition, compared to other species, the scientific community has given less importance to this animal [4]. Thus, these aspects have raised our interest to synthesize previous works on the breeding and feeding of camels in Algeria.

2. Camel Breeding in Algeria

Although camel breeding remains marginal in Algeria, it represents an invaluable animal resource for the desert regions. Thus, the evolution of camel breeding over the past two decades has fluctuated from 234,220 heads in 2000 to 324,199 heads in 2013 and 416,519 in 2018; when considering these numbers against only 120,000 heads in 1987, it shows the importance given to this animal's breeding and its participation in the national economy through the creation of jobs and the satisfaction of the needs for animal products among local populations [5]. Camel breeding in Algeria is distributed over three agro-ecological territories, including the Sahara, the Saharan Atlas and the Steppe. Thus, it is present in 17 Saharan wilayas (prefectures) and 9 steppes [6]. According to a study by Bedda et al. [7], in the region of Ouargla, three systems are identified, including a transhumant system at 77.07%, a sedentary system at 14.01%, and a nomadic system at 8.91%. These results were confirmed by a study carried out by Harek et al. [8], who determined four systems but with different percentages, including nomads at 41%, transhumant people at 21%, semi-nomads at 18%, and sedentary people at 20%. In addition, the color of the coat that falls within the standards of certain animals, such as cattle, is often a characteristic of the breed of this animal. Among the coat colors of camels, Hamra (red dress), which represents 30% of the herd, is appreciated by breeders and considered an animal that tolerates drought and resists disease; Safra (yellow) is frequent at 27.5% and is an animal with one of the favorite colors of farmers; Chegra (light reddish) is found on all farms and throughout the Southwest (11%), but it remains the least appreciated by breeders; Beydha (very light grey or white) is known for its aggressive behavior and susceptibility to disease; Zarga, which is totally black and is often called Zarga, is the least preferred compared to animals with other colors but is not susceptible to diseases; Hadjla, which has a white color on the head and the limbs and a different light yellowish color on the body, is also very appreciated aesthetically, but its frequency is very low in the farms (9.5%); and finally, Zarwala, which has a mixture of blue, white and black (hybrid), remains the least appreciated by breeders and is characterized by a severe form of deafness [9]. In addition, the different populations of camels encountered in Algeria are also distributed according to the geographical area or the tribe to which the animal belongs, namely "Chaambi", "Ouled Sidi Cheikh", "Ait Khebbach", "Camel of the steppe ", "Saharaoui", "Targui or Mehari", "Ajjer", "Reguibi" and "Ftouh" [6].

3. Camel Feeding in Algeria

The diet of the dromedary is based on the different types of Saharan routes and is the main element upon which an extensive camel breeding system is based [10]. However, for a better protection against winter attacks, breeders give them a mixture of wheat, barley and hay in November and release them into the wild in February to enjoy the mountain pastures. The first zone traveled by camels, represented by the steppe and the north of the Sahara (Naama, Bayed, Béchar, Tindouf and the north of Adrar), is characterized by halophilic plants. The farther second zone, which is located south of Adrar and in the extreme south, is characterized by non-halophyte plants. According to a study carried out in Algeria by Slimani et al. [11], plants that are browsed by the dromedary consist mainly of perennial spontaneous plants comprising 13 species belonging to 10 families [10]; in the northern Sahara, there is the presence of 12 species from 8 families. In addition, according to Harek et al. [8], plants that are most browsed by camels with a variable degree of palatability in the Hoggar region are *Tamarix aphylla* and *Tamarix gallica*, as well as other vegetative associations, as shown in Table 1.

Table 1. The different Saharian plant species browsed by camels.

Regions	Plant Species	References
Northern Sahara	*Bassia muricata* (L.) Asch., *Spergularia salina* (Ser.) Presl., *Helianthemum lippii* (L.) Pers., *Argyrolobium uniflorum* Jaub. et Spach., *Astragalus cruciatus* Link., *Astragalus ghyzensis* Bunge., *Lotus roudairei* Bonnet, *Erodium glaucophyllum* (L.) L'Her., *Limoniastrum guyonianum* Boiss., *Cutandia dichotoma* (Forssk.) Trab., and *Zygophyllum album* L.	[10]
Hoggar region	*Panicum turgidum., Balantite aegyptiaca., Cornulaca monocantha., Schowia purpurea., Aristida pengens., Astragalus vogelii., Morettia canescen., Tribulus alatus., Panicum turgidum., Acacia radiana., Acacia seyal., Schowia purpurea., Cornulaca monocantha., Aristida pungens., Moricandia arvensis., Schouwia purpurea., Tribulus terrester., Trichodesma calcaratum., Forsskaolea tenacissima., Maerua crassifolia., Salvodora persica., Atractylis aristata., Balanites aegyptiaca., Echinops bovei., Colocynthis vulgaris.*, and *Atriplex halimus.*	[8]
Sahara	*Launea mucronata., Moltkia ciliata., Oudneya Africana., Pteranthus dichotomus., Helianthemum lipii., Genista saharae., Limoniastrum guyonianum.*, and *Zygophyllum album.*	[11]

4. Conclusions

There have been very few studies conducted on the diet and energy needs of camels in the world and in Algeria. Thus, further studies are recommended in order to determine the needs of camelids in a strategic way. In addition, it would be essential to develop a program involving different countries to measure the energy, protein and other nutrient requirements for camel breeding, growth and racing. Additionally, in-depth studies on genetics would make it possible to determine the standards of camel races in Algeria and in the world.

Author Contributions: Conceptualization, A.L.D.; methodology, A.L.D., S.B. and H.B.; software, A.L.D., S.B. and H.B.; validation, A.L.D.; formal analysis, N.L.; investigation, A.L.D., S.B. and H.B.; data curation, A.B., L.K., I.O., N.S.K., M.L., Z.C. and K.G.; writing—original draft preparation, A.L.D.; writing—review and editing, A.L.D. and N.L.; visualization, E.-H B.; supervision, A.L.D.; project administration, E.-H.B. All authors have read and agreed to the published version of the manuscript.

Funding: This research received no external funding.

Institutional Review Board Statement: Not applicable.

Informed Consent Statement: Not applicable.

Data Availability Statement: The works consulted are detailed in the bibliography.

Acknowledgments: The authors acknowledge the support and help of Mohammed Gagaoua, a researcher from PEGASE INRAE.

Conflicts of Interest: The authors declare no conflict of interest.

References

1. Kadim, I.T.; Purchas, R. Camel carcass and meat quality characteristics. In *More Than Beef, Pork and Chicken–The Production, Processing, and Quality Traits of Other Sources of Meat for Human Diet*; Lorenzo, J., Munekata, P., Barba, F., Toldrá, F., Eds.; Springer: Cham, Switzerland, 2019; pp. 69–94. [CrossRef]
2. Faye, B. How many large camelids in the world? A synthetic analysis of the world camel demographic changes. *Pastoralism* **2020**, *10*, 1–20. [CrossRef]
3. Alaskar, H.M.; Alaqeely, R.; Alhajeri, B.H.; Alhaddad, H. The enigma of camel-types: Localities, utilities, names, and breed statuses. *J. Camelid Sci.* **2021**, *14*, 22–34.
4. Gagaoua, M.; Dib, A.L.; Bererhi, E.-H. Recent Advances in Dromedary Camels and Their Products. *Animals* **2022**, *12*, 162. [CrossRef] [PubMed]
5. Meghelli, I.; Kaouadji, Z.; Yilmaz, O.; Cemal, I.; Karaca, O.; Gaouar, S.B.S. Morphometric characterization and estimating body weight of two Algerian camel breeds using morphometric measurements. *Trop. Anim. Heal. Prod* **2020**, *52*, 2505–2512. [CrossRef] [PubMed]
6. Saadoud, M.; Nefnouf, F.; Hafaoui, F.Z. La viande cameline dans deux régions du Sud Algérien. *Viandes et Produits Carnés* **2019**. Available online: https://www.viandesetproduitscarnes.fr/index.php/fr/1020-la-viande-cameline-dans-deux-regions-du-sudalgerien (accessed on 26 December 2022).

7. Bedda, H.; Adamou, A.; Babelhadj, B. Camel production systems in Algerian Sahara: Case study of the region of Ouargla. *Algerian J. Arid Environ.* **2015**, *5*, 115–127.
8. Harek, D.; El Mokhefi, M.; Ikhlef, H.; Bouhadad, R.; Sahel, H.; Djellout, N.; Arbouche, F. Gene-driving management practices in the dromedary husbandry systems under arid climatic conditions in Algeria. *Pastoralism* **2022**, *12*, 1–12. [CrossRef]
9. Amine, C.Y.; Samir, G.S.B.; Nasreddine, M.; Nacera, T.A.; Nadhira, S.-M. Study of camelina biodiversity in southwestern of Algeria. *J. Life Sci.* **2013**, *7*, 416–427.
10. Trabelsi, H.; Chehma, A.; Al Jassim, R.; Senoussi, A. Camel as seed disperser in the northern Sahara rangelands of Algeria. *Int. J. Biosci.* **2017**, *10*, 58–65. [CrossRef]
11. Slimani, S.; Abdelmadjid, C.; Bernard, F.; Johann, H. Régime et comportement alimentaire du dromadaire dans son milieu naturel désertique en Algérie. *Livest. Res. Rural. Dev.* **2013**, *25*, 213.

biology and life sciences forum

MDPI

Proceeding Paper

Seasonal Variations of Intake in Male Camels on Sahara Range-Lands of Algeria †

Kaouthar Lakhdari *, Tarek Boussaada and Salha Amira Benatallah

Scientific and Technical Research Centre for Arid Areas (CRSTRA), Biskra 07000, Algeria; boussaadatarek@gmail.com (T.B.); benatallahamira@yahoo.fr (S.A.B.)

* Correspondence: kaoutharfido@gmail.com

† Presented at the 10th International Seminar of Veterinary Medicine: Camelids in Algeria & Maghreb, Constantine, Algeria, 20–21 December 2022.

Abstract: The goal of this study was to determine the seasonal and age effects on male camels' intake. Over the two seasons (wet and dry season), follow-ups were conducted at El Alia Rangelands (southeast of Algeria). Eight males were selected to measure bite counts and dry matter intake (DMI) per season; they were distributed into two groups, adult males (AM) and young males (YM). The results showed that there was a significant difference ($p < 0.05$) in the number of bites according to the season and the age categories, with a maximum average of 80.33 ± 18.206 for AM in the dry season and a minimum average of 15.50 ± 9.955 for YM in the wet season. During the wet season, *Traganum nudatum* had the highest dry matter intake (DMI), with 2.01 kg DM for AM and 0.28 kg DM for YM, while *Salsola longifolia* and *Salsola tetragona* had the lowest, with 0.38 and 0.39 kg DM, respectively, during the dry season in YM. *Limoniastrum guyonianum* exhibited high amounts of dry matter intake (DMI) in the dry season, with 1.10 and 0.22 kg DM in AM and YM, respectively, but significantly decreased in AM and YM during the wet season (0.03 and 0.02 kg DM respectively). The season influences the feed intake of male camels, of course.

Keywords: camel; feed intake; sahara rangeland; season; vegetation

Citation: Lakhdari, K.; Boussaada, T.; Benatallah, S.A. Seasonal Variations of Intake in Male Camels on Sahara Range-Lands of Algeria. *Biol. Life Sci. Forum* **2023**, *22*, 14. https://doi.org/10.3390/blsf2023022014

Academic Editors: Amira Leila Dib, Said Boukhechem, Hithem Bougherara and El-Hacene Bererhi

Published: 7 April 2023

1. Introduction

Under climate change, the dromedary is a vital animal for the Algerian population's food security. Camels are a source of meat and milk protein of great value. In particular, in arid areas, camel breeding contributes to food security related to livestock production, improving the living conditions of pastoralists, improving farmers' income, and maintaining biodiversity by fair use of spontaneous feed by camel herds [1]. Despite this position, the implemented research and development programs did not grant a fair evaluation, which is why one of the issues deemed urgent is being able to permanently determine all the indicators of camel breeding, particularly those relating to rangeland feeding. The food of camels seems unknown until this day; indeed, there are few studies of dromedary palatability in Algeria [2]. The dromedary diet remains one of the least-studied aspects of this species. In order to predict the dromedary's effect on the vegetation and its nutrient requirements, it is important to understand the dromedary's foraging behavior. Furthermore, managing and utilizing range plants in camel-feeding systems necessitates a thorough understanding of seasonal variation in intake. The aim of the study was to determine the impact of seasonal variation on adult camels' intake in order to satisfy their dry matter and nutrient requirements and to improve animal and housing management.

2. Materials & Methods

Thirty-two male camels were randomly selected, marked, and followed. During the wet and dry seasons, eight animals from each category (adult males (AM) and young males (YM)) were observed for two hours, one animal per day and category. The bite count technique is used to quantify forage consumption [3]. Forage intake was estimated after cutting and weighing simulated bites that represent ingested bites by multiplying bite counts by the average mass per bite [4]. The intake of dry matter (DMI) is determined by the following formula: DMI = GT/R × Σ (TNBi × WBi).

3. Results and Discussion

The results obtained (Figure 1) showed that for AM, the number of bite counts ranged from 4 to 89 in the wet season and from 57 to 102 in the dry season. The bite weight ranged from 0.76 g for *Anabasis articulate* in the dry season to 3.77 g recorded for *Traganum nudatum* in the wet season. For YM, the number of bite counts in the wet season was 7 to 29 and 8 to 48 in the dry season, respectively. The bite weights ranged from 0.57 g for *Anabasis articulata* during the wet season and 1.61 g measured for *Traganum nudatum* in the dry season. Overall, there is a significant difference ($p < 0.05$) in the average number of bites between seasons, with the dry season having a higher average number of bites than the wet season (33.33 ± 3.572 and 15.50 ± 9.955, respectively) for both (AM) and (AM) (80.33 ± 18.206 and 30.83 ± 14.959, respectively). The preferred species by male camels display a variation in the number of bites ($p < 0.05$); it is obvious that, during the dry season, *Limoniastrum guyonianum*, which is neglected during the wet season, is preferred. According to Ref. [5], this variability can be justified by the smell of this species during winter; it is the sense of smell that guides the animal in its choice more than the taste and view, while *Traganum nudatum* in both seasons remains very grazed. The total number of bites is inversely proportional to the bite's weight [6]. The number of bites exhibits a significant difference between AM and YM ($p < 0.05$). Regarding the amount of dry matter intake (DMI) (Figure 2), for AM the highest quantity during the wet season is recorded for *Traganum nudatum* 2.01 kg DM, but *Limoniastrum guyonianum* recorded the highest amount with 1.1 kg MS during the dry season. For (YM), *Traganum nudatum* remains the highest daily intake of dry matter recorded in the wet and dry seasons, 0.28 and 0.36 kg DM, respectively. The study demonstrated a difference in the intake of dry matter between seasons ($p < 0.05$); this result is confirmed by Ref. [7]. Plant intake depends on their feeding behavior.

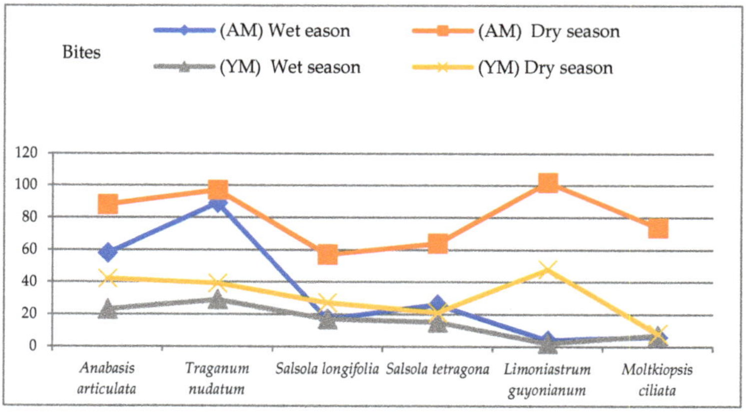

Figure 1. The number of bites by an adult and young males.

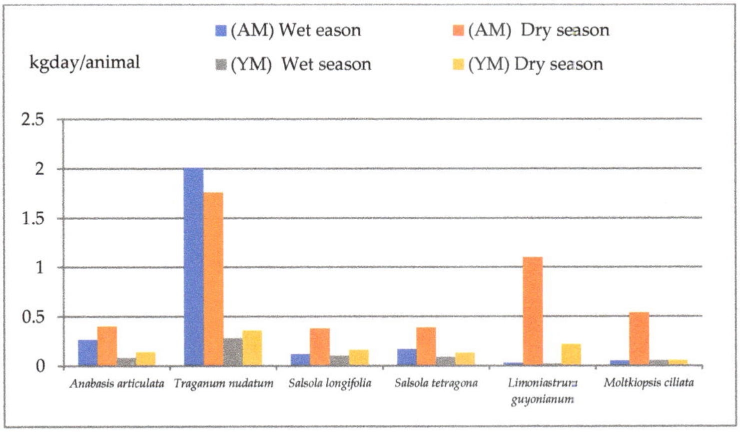

Figure 2. Dry matter intake by an adult and young males.

4. Conclusions

The use of each species by camels varies seasonally; this nutritional instinct has a positive ecological impact on the preservation of the vegetation cover of saharan rangelands.

Author Contributions: Conceptualization, K.L. and S.A.B.; methodology, K.L.; software, T.B.; validation, K.L., S.A.B. and T.B.; formal analysis, T.B.; investigation, K.L.; resources, K.L.; data curation, K.L.; writing—original draft preparation, K.L. and S.A.B.; writing—review and editing, K.L., S.A.B. and T.B.; visualization, T.B.; supervision, K.L.; project administration, K.L.; funding acquisition, K.L. All authors have read and agreed to the published version of the manuscript.

Funding: This research received no external funding.

Institutional Review Board Statement: Not applicable.

Informed Consent Statement: Not applicable.

Data Availability Statement: The data presented in this study are available on request from the corresponding author.

Acknowledgments: The authors are thankful to all the camel herders in the El Alia region southeast of Algeria, who allowed us to follow up on their animals.

Conflicts of Interest: The authors declare no conflict of interest.

References

1. Lakhdari, K.; Belhamra, M.; Mayouf, R. Forage species preferred by dromedaries and their chemical composition in arid rangelands of Algeria. *Livest. Res. Rural Dev.* **2015**, *27*, 10.
2. M'hammed, B. Etude Floristique et Nutritive Spatio-Temporelle des Parcours Camelins du Sahara Occidental Algerien: Cas des Régions de Bechar et Tindouf. Ph.D. Thesis, Université Kasdi Merbah, Ouargla, Algeria, June 2013. (In French)
3. Gauthier-Pilters, H. Observations on the ecology of the dromedary in North Western Sahara. *Mammalia* **1961**, *25*, 195–280.
4. Dumont, B. Préférences et sélection alimentaire au pâturage. *INRA Prod. Anim.* **1996**, *9*, 359–365. (In French) [CrossRef]
5. Thewis, A.; Bourbouze, A.; Compère, R.; Duplan, J.-M.; Hardouin, J. *Manuel de Zootechnie Comparée Nord-Sud*; INRA Editions: Paris, France, 2005. (In French)
6. Mebirouk-Boudechiche, L.; Boudechiche, L.A.; Ferhat, R.; Tahar, A. Relationship between availability of grass, food intake and grazing activities of sheep. *Arch. Zootec.* **2014**, *63*, 277–287. [CrossRef]
7. Kamoun, M.; Steinmetz, P. Feeding behaviour, intake and digestion of the Camelus dromedarius at pasture. *Opt. Méditerr. Sér. B Etudes Rech.* **1995**, *13*, 51–57.

biology and life sciences forum

MDPI

Proceeding Paper

Valorization of Camel Meat and Meat Products in the World and in Algeria †

Hithem Bougherara [1], Amira Leila Dib [1,*], Said Boukhechem [1], Assia Bouaziz [1], Louiza Kadja [1], Khireddine Ghougal [1], Ibtissem Oubira [1], Zahia Chari [1], Nessrine Sana Kerrour [2], Melisa Lamri [3], Nedjoua Lakhdara [1] and El-Hacene Bererhi [1]

1 Gestion Santé et Productions Animales Research Laboratory, Institut des Sciences Vétérinaires El-Khroub, Université Frères Mentouri Constantine 1, Constantine 25000, Algeria; heithem1985@yahoo.fr (H.B.); said.boukhechem@umc.edu.dz (S.B.); bouaziz.assialp@gmail.com (A.B.); louiza20132014@gmail.com (L.K.); ghougal29@gmail.com (K.G.); ibtissem.oubira@yahoo.fr (I.O.); manel_kaouther25@yahoo.fr (Z.C.); nedjoua2002@hotmail.com (N.L.); brerhihacene@yahoo.fr (E.-H.B.)
2 Department of Food Science, Université Abderrahmane Mira, Béjaïa 06000, Algeria; sanakerrour@outlook.fr
3 Department of Food Science, Mouloud Mammery University, Tizi-Ouzou 15000, Algeria; lamrimeliza1@gmail.com
* Correspondence: dib.amiraleila@umc.edu.dz

† Presented at the 10th International Seminar of Veterinary Medicine: Camelids in Algeria & Maghreb, Constantine, Algeria, 20–21 December 2022.

Abstract: Camel meat is considered as good source of nutrition, its taste and texture are similar to that of cattle and has an amino acid content ten times higher than that of the latter. Moreover, its consumption is very low and the manufacture and marketing of cameline meat products are very neglected. The objective of this work is to promote camel's meat and the derived products from it. Thus, the industry sector should launch and invest in the production of large ranges of camel meat, in order to meet consumer demands.

Keywords: camelids; Algeria; meat; meat products; consumers

Citation: Bougherara, H.; Dib, A.L.; Boukhechem, S.; Bouaziz, A.; Kadja, L.; Ghougal, K.; Oubira, I.; Chari, Z.; Kerrour, N.S.; Lamri, M.; et al. Valorization of Camel Meat and Meat Products in the World and in Algeria. *Biol. Life Sci. Forum* **2023**, 22, 11. https://doi.org/10.3390/blsf2023022011

Academic Editor: Boussena Sabrina

Published: 4 April 2023

1. Introduction

Food security and nutrition are major issues on the global agenda, linked to the Sustainable Development Goals (SDGs), which among other things, aims to end all forms of malnutrition in the world by 2030 [1]. In addition, the world population is increasing every year with a steady increase in the demand for food in the world, which makes various countries face acute food crises [2]. Thus, it is possible to meet these requirements as a result of the consumption and creation of combined products using animal and vegetable raw materials enriched with certain vitamins and biologically active additives and whose wholesomeness lies in the balance and improvement of the diet, thanks to the introduction of proteins, amino acids, vitamins, micro and macro elements, dietary fiber and other beneficial substances. Among the main sources of nutrients necessary for maintaining the normal functioning of the human body, meat and its derivatives [3]. Meat is defined as the whole carcass of cattle, sheep, goats, camels, buffaloes, deer, hares, poultry or rabbits and is an important source of nutrients, such as protein, iron and vitamins and has beneficial effects on human health [4]. Its production is estimated at around 340 million tons per year and the average annual consumption is estimated at 43 kg of meat/person, with a higher rate in developed countries [5]. Thus, camels can play a very important role by providing an important part of the human diet in particular, meat, in order to meet the demand [6]. All these aspects have aroused the interest of synthesizing the various research and works valuing meat and meat products from camels.

2. Camel Meat Consumption and Meat Products in the World and in Algeria

Although consumers are often surprised to learn that camel meat is consumable. Indeed, it is considered as an excellent source of protein with many medicinal benefits for human health. However, its contribution to world meat production is rather marginal due to the minor importance of camels among herbivores. Indeed, compared to all types of meat producers except fish, this meat represents only 0.13% of the total meat produced in the world and 0.45% of the red meat of herbivores. On the other hand, this contribution must be evaluated especially in arid zones [7]. In fact, camel meat is very important in arid and semi-arid regions of the world and it constitutes an important resource for some countries such as Sudan, Somalia and Mauritania [8]. In addition, regarding healthy and nutritional value, camel's meat contains 78% water, 19–22% protein and 3% fat and could be a worthy choice as it contains high quality proteins and produces carcasses with less fat, fewer calories and less cholesterol than other animal meats [9]. Every year, around 250,000,000 camels are slaughtered in different countries. Africa produces 62.2% of the world's camelid meat, followed by Asia at 35.8%. While South America only contributes 5.3%. In Africa, the largest contribution is from East Africa, followed by North Africa and West Africa. Furthermore, the production of camel meat is probably underestimated because a large number of camels are slaughtered outside official circuits and are therefore not included in the statistics [10]. Its consumption in North African countries is around 2.84 kg/capita/year [11], while, it is very low in Algeria, although it represents 33% of red meat slaughtering and this percentage is constantly changing [12]. In fact, during the period from 2012–2015, the slaughter of the camel species in Tamanrasset and Ghardaïa (Algeria), represents respectively 17% and 22% of the total meat produced [13]. Moreover, the availability of camel meat in the different regions of Algeria, in the South in the comparison with the North of the country, could influence its consumption frequency. A recent study reported by Lamri et al. [14], using an online survey (n = 665 consumers), showed that camel meat is not well appreciated and eaten by consumers from the region of Kabylia (Algeria), 54.3% of them never consumed it, only 1.6% eat it always, 35% sometimes and 9.1% rarely. These results may be due to habits and traditions and the limit production and availability of camel meat in this region and in the North of the country in general.

In addition, camel meat is rarely processed and its by-products remain underdeveloped. However, fermented or unfermented drying, brining and smoking of cameline meat are common traditional practices, especially in Algeria and Morocco [15]. Indeed, in some countries in order to promote this species, cameline meat is transformed into hamburgers, patties, sausages, cashir [8,16–19].

3. Conclusion

Despite the low productivity of camel meat, particularly outside arid and semi-arid regions. This species as well as its meats and meat products constituting some food rich in animal proteins, should be more valued in all regions and countries, which will allow the most disadvantaged populations to have access to red meats at a lower price. Moreover, the processing of cameline meats is neglected all over the world and in order to constitute a wide range of camel products, butcher shops and processing industries should be set up.

Author Contributions: Conceptualization, A.L.D.; methodology, A.L.D., S.B. and H.B.; software, A.L.D., S.B. and H.B.; validation, A.L.D.; formal analysis, N.L.; investigation, A.L.D., S.B. and H.B.; data curation, A.B., L.K., I.O., N.S.K., M.L., Z.C. and K.G.; writing—original draft preparation, A.L.D.; writing—review and editing, A.L.D. and N.L.; visualization, E.-H.B.; supervision, A.L.D.; project administration, E.-H.B. All authors have read and agreed to the published version of the manuscript.

Funding: This research received no external funding.

Institutional Review Board Statement: Not applicable.

Informed Consent Statement: Not applicable.

Data Availability Statement: Not applicable.

Acknowledgments: The authors acknowledge the support and help of Mohammed Gagaoua, a researcher from PEGASE INRAE.

Conflicts of Interest: The authors declare no conflict of interest.

References

1. Orazov, A.; Nadtochii, L.; Bozymov, K.; Muradova, M.; Zhumayeva, A. Role of Camel Husbandry in Food Security of the Republic of Kazakhstan. *Agriculture* **2021**, *11*, 614. [CrossRef]
2. Al-Mahish, M.; Elzaki, R.; Al-Qahtani, N. Demand and nutrients elasticities of camel meat: An analysis of kingdom of Saudi Arabia. *Livest. Sci.* **2018**, *9*, 140–150.
3. Kaimbayeva, L.; Kenenbay, S.; Dikhanbaeva, F.; Tnymbaeva, B.; Kazihanova, S. Histological studies of the muscle tissue of the bactrian camel meat in the process of autolysis. *Food Sci. Technol.* **2020**, *41*, 371–375. [CrossRef]
4. Gagaoua, M.; Picard, B. Current advances in meat nutritional, sensory and physical quality improvement. *Foods* **2020**, *9*, 321. [CrossRef] [PubMed]
5. Mohammed, S.A.A. A study of cholesterol concentrations of camel meat and beef. *Int. J. Agric. Innov. Res.* **2019**, *7*, 397401.
6. Osaili, T.M.; Hasan, F.; Al-Nabulsi, A.A.; Dhanasekaran, D.K.; Obaid, R.S.; Hashim, M.S.; Radwan, H.M.; Cheikh Ismail, L.; Hasan, H.; Faris, M.A.-I.E.; et al. Effect of Essential Oils and Vacuum Packaging on Spoilage-Causing Microorganisms of Marinated Camel Meat during Storage. *Foods* **2021**, *10*, 2980. [CrossRef] [PubMed]
7. El-Badawi, A. The present situation of animal protein in Egypt and the role of camels in providing cheap and healthy meat for people in poor greenery lands. *Int. J. Avian Wildl. Biol.* **2018**, *3*, 319–322.
8. Faye, B.; Abdelhadi, O.; Raiymbek, G.; Kadim, I.; Hocquette, J.-F. La production de viande de chameau: état des connaissances, situation actuelle et perspectives. *INRA Prod. Anim.* **2013**, *26*, 289–299. (In French) [CrossRef]
9. Abdel-Raheem, H.; Ahmed, H.; Abd-Allah, S.; Abdel-Rasoul, M. Nutritive value of the Dromedary camel meat. *SVU-Int. J. Vet. Sci.* **2019**, *2*, 68–74. [CrossRef]
10. Kadim, I.T.; Mahgoub, O.; Faye, B.; Farouk, M.M. *Camel Meat and Meat Products*; CABI: Wallingford, UK, 2012.
11. Faye, B.; Jaouad, M.; Bhrawi, K.; Senoussi, A.; Bengoumi, M. Elevage camelin en Afrique du Nord: état des lieux et perspectives. *Rev. Elev. Med. Vet. Pays Trop.* **2014**, *67*, 213–221. (In French) [CrossRef]
12. Benaissa, A.; Ould El Hadj-Khelil, A.; Adamou, A.; Babelhadj, B.; Hammoudi, M.; Riad, A. Qualité de la viande de dromadaire dans les abattoirs de Ouargla en Algérie. II. Contamination bactérienne superficielle des carcasses. *Rev. Elev. Med. Vet. Pays Trop.* **2014**, *67*, 229–233. (In French) [CrossRef]
13. Sadoud, M.; Nefnouf, F.; Hafaoui, F. La viande cameline dans deux régions du Sud Algérien: La place de l'élevage, de la transformation et de la consommation de la viande cameline dans les deux régions algériennes Tamanrasset et Ghardaïa. *Viandes Prod. Carnés* **2019**, *35*, 3. (In French)
14. Lamri, M.; Djenane, D.; Gagaoua, M. Goat meat consumption patterns and preferences in three provinces of Kabylia region in Algeria compared to other meat species: Results of an online survey. *Meat Technol.* **2022**, *63*, 96–108. [CrossRef]
15. Gagaoua, M.; Boudechicha, H.-R. Ethnic meat products of the North African and Mediterranean countries: An overview. *J. Ethn. Foods* **2018**, *5*, 83–98. [CrossRef]
16. Zidan, M.; Kassem, A.; Dougbag, A.; Ghazzawi, E.E.; El Aziz, M.A.; Pabst, R. The spleen of the one humped camel (*Camelus dromedarius*) has a unique histological structure. *J. Anat.* **2000**, *196*, 425–432. [CrossRef] [PubMed]
17. Boudechicha, H.-R.; Sellama, M.; Lamri, M.; Boudjellal, A.; Gagaoua, M. Produits carnés traditionnels des pays d'Afrique du Nord. *Viandes Prod. Carnés.* **2018**, *34*, 1–19. (In French)
18. Zaki, E.F. The quality characteristics of camel sausage formulated with different levels of whey protein powder. *Int. J. Agric. Environ. Biotechnol.* **2017**, *2*, 238929.
19. Zakaria, B.; Abdelhakim, S.; Bernard, F. Camel meat marketing and camel meat marketplace in the Algerian northern Sahara-case of the region of Souf. *Emir. J. Food Agric.* **2020**, *32*, 319–327. [CrossRef]

biology and life sciences forum

MDPI

Proceeding Paper

Dried Camel (*Camelus dromedarius*) Meat Contributing to Food Safety †

Youcef Rahmani [1,2,3,*] and Réda Khama [2,4]

1 Department of Process Engineering, University of Ouargla, Ouargla 30000, Algeria
2 Laboratory of Process Engineering, Faculty of Applied Sciences, University of Ouargla, Ouargla 30000, Algeria; r.khama@univ-batna2.dz
3 Scientific and Technical Research Centre in Physico-Chemical Analysis, Bousmail 42004, Algeria
4 Department of Process Engineering, Faculty of Technology, University of Batna 2, Batna 05000, Algeria
* Correspondence: rahmani.youcef@univ-ouargla.dz

† Presented at the 10th International Seminar of Veterinary Medicine: Camelids in Algeria & Maghreb, Constantine, Algeria, 20–21 December 2022.

Abstract: The effect of static drying on the physicochemical quality of camel (*Camelus dromedarius*) meat slices as driedin an oven at 65 °C, with the aim of contributing to food safety. During the experiments, meat was cutinto. 8 ± 0.2 cm thick slices, soaked in a saline solution for 30 min, and then dried. Moisture content wasmeasured during the drying phase, and physicochemical parameters, such as protein, ash, sodium, pH, and mineral content, such as calcium (Ca), phosphor (P), magnesium (Mg), potassium (K), zinc (Zn), iron (Fe), copper (Cu) and manganese (Mn), were monitored before and after drying. The colorimetric parametersof the dried samples are also measured. The results show that the physicochemical quality of the dried camelmeat meets the requirements of the Codex Alimentarius Commission. Indeed, the moisture content decreased from 73.94 ± 0.51% to 13.33 ± 0.44%, and the dried food samples were characterized by increases (i) from 19.72 ± 0.30% to 50.97 ± 0.65% in protein content; (ii) from 1.115 ± 0.012% to 4.781 ± 0.047%, in ash content; (iii) from 260 ± 11.7 mg to 1690 ± 32 mg, in the sodium content (for 100 g of dry matter), and (iv) from 5.956 ± 0.087 to 6.203 ± 0.091 in pH value. However, the variation is not significant for all mineral content parameters. Finally, themean values of the colorimetric parameters of brightness (L*), redness (a*), and yellowness (b*) are 37.13 ± 1.64, 22.02 ± 0.72, and 7.73 ± 0.69, respectively (before drying) and they are 25.57 ± 1.56, 9.43 ± 0.78, and 3.74 ± 0.21, respectively (after drying).

Keywords: *Camelus dromedarius*; camel meat; drying; quality; safety food

Citation: Rahmani, Y.; Khama, R. Dried Camel (*Camelus dromedarius*) Meat Contributing to Food Safety. *Biol. Life Sci. Forum* **2023**, 22, 18. https://doi.org/10.3390/blsf2023022018

Academic Editors: Amira Leila Dib, Said Boukhechem, Hithem Bougherara and El-Hacene Bererhi

Published: 11 April 2023

1. Introduction

The dromedary (*Camelus dromedarius*) is a species of domestic mammal of the family Camelidae and genus Camelus [1]. The exploitation of the productive potential of the livestock sector is based on several sectors, the main ones being meat, milk, and its derivatives [2]. The consumption of camel meat is one of the main resources for the population of arid areas, despite itscontribution to the world's production of red meat being very marginal. Several traditional processing methods and preservation are applied for red meat, salted undried, smoked, dried unfermented, fermented semi-dried or dried, cooked, or preservedin fat.

In order to keep the final quality of the processed product [3], several physicochemical parameters are checked during the drying process in order to have a product that meets the national and international requirements of dried meat quality that can be preserved for several months and would beappreciated by caravanners and nomads during their travels. Moreover, dried camel meat is still sold in various preparations, such as sausages, in supermarkets [4,5]. The increase in camel consumption and production has highlighted the role of camel meat in the diet of the population in arid regions [6]. Dried camel meat is a primary foodsource in arid regions, which is also part of the traditionsofthe population.Indeed, some authors and investigators in the field of consumption estimate that a carcass of an adult dromedary is likely to provide 40 kg of bones, 16 kg of meat, and 10 kg of fat, which can meet an adult man's energy needs for up to 5 days and up to 35 days of protein requirements. From a nutritional point of view, camel meat contains between 20% and 23% protein.

The aim of this experimental investigation is to prove that the drying process of camel meat does not affect its physicochemical quality. On the contrary, drying is a good method forincreasing the shelf life of foods, and it could contribute to food security.

2. Materials and Methods

To carry out this work, we proceeded to a sampling themeat of young camels (less than 4 years old) coming from the regional slaughterhouse of Ouargla City, Algeria. It was thinly sliced (0.8 ± 0.2 cm thickness; 10 ± 0.2 cm length), soaked for 30 min in a cold saline solution of 19% salinity, then dried at 65 °C in an oven. The physicochemical analysis is carried out according to the ISO standards relating to the meat analysis, before and after drying, with both fresh and dried samples.

3. Results and Discussion

The application of the methods of physicochemical analysis relating to meat and meat products on fresh and dried camel meat after salting gives the results summarized in Tables 1 and 2. The increase in dry matter is due to the drying effect caused by loss of water content [7], and the change in color, particularly the parameters a* and b*, reflect the oxidation of myoglobin and make the meat darker. Finally, the increase in sodium content is due to salting.

Table 1. Physicochemical quality of the fresh and dried camel meat.

Composition	Fresh Camel Meat	Camel Meat Dried at 65 °C
Moisture content %	73.94 ± 0.51	13.33 ± 0.44
Dry matter %	26.06 ± 0.47	86.67 ± 0.34
Ash %	1.115 ± 0.012	4.781 ± 0.047
Protein %	19.72 ± 0.30	50.97 ± 0.65
Lipid %	3.744 ± 0.14	3.19 ± 0.29
Sel %	-	3.63 ± 0.21
pH	5.956 ± 0.087	6.203 ± 0.091
Color L*, a*, b*	37.13 ± 1.64, 22.02 ± 0.72, 7.73 ± 0.69	25.57 ± 1.56, 9.43 ± 0.78, 3.74 ± 0.21

Table 2. Minerals and nutritional composition of the fresh and dried camel meat.

Composition (mg/100 g dm)	Fresh Camel Meat	Camel Meat Dried at 65 °C	Composition (ug/100 g dm)	Fresh Camel Meat	Camel Meat Dried at 65 °C
Ca	58.3 ± 6.4	56.2 ± 3.7	Zn	12,830 ± 649	11,983 ± 560
P	675.7 ± 23.6	670.5 ± 13.9	Fe	6647 ± 944	6821 ± 560
Mg	917 ± 83	903 ± 58	Cu	1025 ± 30	1043 ± 22
Na	260 ± 11.7	1690 ± 13	Mn	460 ± 230	390 ± 190
K	70 ± 8.7	64 ± 7.4	-	-	-

4. Conclusions

The production and consumption of camel meat can contribute to food security, in particular, as a source of protein in arid regions and for caravans and nomads. The drying process could be an additionalmethod for processing and preserving cameline meat, and the nutritional value and physicochemical quality of the dried meat meet the required standards of the food code.

Author Contributions: Conceptualization, Y.R. and R.K.; methodology, Y.R. and R.K.; software, Y.R.; validation, Y.R.; formal analysis, Y.R. and R.K.; investigation, Y.R.; resources, Y.R.; data curation, Y.R.; writing—original draft preparation, Y.R.; writing—review and editing, R.K.; visualization, R.K. and Y.R.; supervision, R.K.; project administration, R.K.; funding acquisition, Y.R. All authors have read and agreed to the published version of the manuscript.

Funding: This research received no external funding.

Institutional Review Board Statement: Not applicable.

Informed Consent Statement: Not applicable.

Data Availability Statement: The works consulted are detailed in the bibliography.

Acknowledgments: The authors thank the members of the Ecosystem Laboratory at the University of Ouargla, as well as the person in charge of the Process Engineering Laboratory (LGP) and the personnel of the Research Center and Physicochemical Analyses (CRAPC) for their support with the materials used inthis work.

Conflicts of Interest: The authors declare no conflict of interest.

References

1. Baz, Y.; Khomeiri, M. Introduction to Camel origin, history, raising, characteristics, and wool, hair and skin: A Review. *Res. J. Agric. Environ. Manag.* **2015**, *4*, 496–508.
2. Cherifi, Y.A.; Gaouar, S.B.S.; Moussi, N.; TabetAoul, N.; Saïdi-Mehtar, N. Study of Camelina Biodiversity in Southwestern of Algeria. *J. Life Sci.* **2013**, *7*, 416–427.
3. Faye, B. The Camel Today: Assets and Potentials. *Anthropozoologica* **2014**, *49*, 167–176. [CrossRef]
4. Rahmani, Y.; Khama, R. Study the effect of sun and microwave drying on quality of camel meat (camelus dromedarius) slices. *Alger. J. Arid Environ.* **2022**, *12*, 98–107.
5. Sahraoui, N.; Dotreppe, O.; Brahim Errahmani, M.B.; Boudjenah, S.; Baaissa, B.; Guetarni, D.; Hornick, J.-L. Caractérisation des acides gras de la viande cameline en Algérie. *Cah. Nutr. Diet.* **2014**, *49*, 231–234. [CrossRef]
6. Suliman, G.M.; Alowaimer, A.N.; Hussein, E.O.S.; Ali, H.S.; Abdelnour, S.A.; Abd El-Hack, M.E.; Swelum, A.A. Chemical Composition and Quality Characteristics of Meat in Three OneHumped Camel (*Camelus dromedarius*) Breeds as Affected by Muscle Type and Post-Mortem Storage Period. *Animals* **2019**, *9*, 834. [CrossRef]
7. Rahmani, Y.; Khama, R. Effect of thickness and pretreatment on the physicochemical quality of sun-dried camelina meat. In Proceedings of the 2020 6th International Symposium on New and Renewable Energy (SIENR), Ghadaia, Algeria, 13–14 October 2021; IEEE: New York, NY, USA; pp. 1–5.

biology and life sciences forum

MDPI

Proceeding Paper

The Use of a Natural Product "Camel Milk" as a Regulator of Glycemia †

Louiza Kadja *, Assia Bouaziz and Amira Leila Dib

Gestion Santé et Productions Animales Research Laboratory, Institut des Sciences Vétérinaires El-Khroub, Univesité Frères Mentouri Constantine 1, Constantine 25000, Algeria; bouazizassialp@gmail.com (A.B.); dib.amiraleila@umc.edu.dz (A.L.D.)

* Correspondence: louiza20132014@gmail.com

† Presented at the 10th International Seminar of Veterinary Medicine: Camelids in Algeria & Maghreb, Constantine, Algeria, 20–21 December 2022.

Abstract: In many countries around the world, camel milk is a healthy food used to treat many health issues including diabetes. Thus, it has been demonstrated from several studies in vivo that the consumption of camel milk either fresh or fermented could have a positive effect on certain diseases and metabolic disorders such as hypercholesterolemia and hypertension. Furthermore, lactic cultures from camel milk have also been tested for the action of some active proteins or probiotic and have shown encouraging results. The objective of this work is to synthesize the data related to the benefits of camel milk and its effect on the regulation of glycemia. More in-depth studies should be carried out on humans, in order to confirm the effect of camel milk on glycemia.

Keywords: milk; camel; diabetes; proteins; probiotics; health

Citation: Kadja, L.; Bouaziz, A.; Dib, A.L. The Use of a Natural Product "Camel Milk" as a Regulator of Glycemia. *Biol. Life Sci. Forum* **2023**, 22, 22. https://doi.org/10.3390/blsf2023022022

Academic Editors: Said Boukhechem, Hithem Bougherara and El-Hacene Bererhi

Published: 13 April 2023

1. Introduction

In recent decades, diabetes has been a major public health problem worldwide, with a prevalence that could reach 700 million by 2045 [1]. In addition, type 2 diabetes, which accounts for approximately 90% of diabetes cases, can lead to serious damage to the heart, eyes, kidneys, blood vessels and nerves [2]. As a result, many studies have focused on the development of anti-diabetic drugs and functional foods to cure or minimize this damage [3,4]. In the arid regions of Africa and Asia, camel milk can provide to the nutritional needs of these minor populations. Furthermore, it is recommended in these regions to consume this milk in a fresh or fermented state for the treatment of diabetes [5]. Indeed, it has been recently reported that camel milk may have medicinal properties [6], such as anticarcinogenic, antimicrobial, antioxidant, angiotensin I converting enzyme inhibitory activities, as well as cholesterol-lowering, hypoglycemic and hypoallergenic effects due to the presence of bioactive compounds [5]. In addition, the fermentation of camel milk by beneficial microorganisms offers consumers, in addition to good nutritional value, prevention against diabetes because they can reduce the absorption of glucose in the intestines. They are therefore considered one of the best ways to manage high blood sugar [7]. The objective of this work is to demonstrate the involvement of camel milk, active proteins or probiotics isolated from it, in the regulation of blood sugar and its possible use in the control of diabetes.

2. Materials and Methods

A search for articles was carried out using both Google Scholar and NCBI PubMed databases. All the articles uploaded focused on the regulation glycemia effect of camel milk and active proteins or probiotics derived from it.

3. Results and Discussion

Table 1 represents the analysis of data carried out from some selected papers and relating to the hypoglycemic effect of camel milk.

The effect of camel milk and its derivative products on the various parameters of diabetes has been reported by numerous studies cited in the table below. Zheng et al. [8] reported that the hypoglycemic effect of this milk is due to an insulin-like protein. Further, Kilari et al [9] demonstrated that a camel milk protein hydrolysates, source of bioactive peptides, could activate the insulin receptor and prevent hyperglycemia and diabetes complications. On the other hand, some authors such as Manaer et al. [10] and Chouikhi et al. [5] have suggested that the hypoglycemic effect of camel milk is mainly linked to its richness in probiotics. Overall, however, the number of these studies using probiotics isolated from camel milk remains relatively low compared to studies that used either raw or fermented milk.

Table 1. Summary of results.

References	Model of Study	Diabetogenic	Products and Dose/Day	Duration	Strains of Probiotic	Benefic Effect on Diabetes Parameters
Agrawal et al. [1]	Rats	Streptozotocin	Fresh camel milk (250 mL)	3 weeks	N.D	↓ Glycemia
El-Said et al. [7]	Rabbits	Alloxan	Fresh camel milk (7 mL/kg)	4 weeks	N.D	↓ Glycemia + ↑ Insulinemia
Alharbi et al. [2]	Rats	Streptozotocin	Fermented camel milk (5 mL)	28 days	N.D	↓ Glycemia + Hepatoporotector effect
Fallah et al. [11]	Patients	-	Fermented camel milk (250 mL)	8 weeks	N.D	↑ Insulinemia
Xu et al. [12]	Mice	LPS/D-GalN	Probiotic isolated from Mongolian camel milk	7 weeks	*Lactobacillus Paracasei* subsp. *paracasei* WXD5	↓ lipopolysacharids + ↓ IL6
Chouikhi et al. [5]	Rats	Alloxan	Probiotic isolated from Tunisian camel milk (10^9 cfu/mL)	14 days	*Lactiplantibacillus plantarum* LC38	Hepatoprotector effect
Manaer et al. [10]	Mice	*db/db mice*	Traditional fermented cheese whey (10^8–10^{10} cfu/mL)	6 weeks	*Lactobacillus Kefiranofaciens* + *Issatchenkia orientalis*	↓ Glycemia, OGTT, HbAlc

↑: Increase; ↓: decrease; N.D: no determined; OGTT: oral glucose tolerance test; HBAlc: Hemoglobin A1c; IL6: Interleukin 6.

4. Conclusions

Most of the results of in vivo tests searching the effect of camel milk, have shown that this animal product could have preventive and/or curative effects against diabetes, due to the action of some active proteins. Moreover, very little research have been done, on the strains of probiotics isolated from this milk and their effects on diabetes. Thus, other trials on the different strains isolated from camel milk and tested on several animal species as well as on humans, should be considered.

Author Contributions: Conceptualization, L.K. and A.L.D.; methodology, L.K.; software, L.K.; validation, A.L.D., A.B.; formal analysis, L.K.; investigation, L.K.; resources, L.K.; data curation, A.L.D.; writing—original draft preparation, L.K.; A.L.D. and A.B.; writing—review and editing, L.K., A.B. and A.L.D.; visualization, L.K., A.L.D.; supervision, A.L.D.; project administration, A.L.D.; funding acquisition, L.K. All authors have read and agreed to the published version of the manuscript.

Funding: This research received no external funding.

Institutional Review Board Statement: Not applicable.

Informed Consent Statement: Not applicable.

Data Availability Statement: Not applicable.

Acknowledgments: This work is a part of the "probiotic" project agreement number D01N01UN25012 0180004. The authors acknowledge the support and help of El-Hacene Bererhi, the Director of the Institute of Veterinary Sciences, Université Frères Mentouri Constantine 1, Algeria; Mohammed Gagaoua, a researcher from PEGASE INRAE, France and Nedjoua Lakhdara, a member of GSPA Research Laboratory, Institute of Veterinary Sciences, Université Frères Mentouri Constantine 1, Algeria.

Conflicts of Interest: The authors declare no conflict of interest.

References

1. Agrawal, R.; Kochar, D.; Sahani, M.; Tuteja, F.; Ghorui, S. Hypoglycemic activity of camel milk in streptozotocin induced diabetic rats. *Int. J. Diabetes Dev. Ctries.* **2004**, *24*, 47–49.
2. Alharbi, Y.M.; Sakr, S.S.; Albarrak, S.M.; Almundarij, T.I.; Barakat, H.; Hassan, M.F.Y. Antioxidative, Antidiabetic, and Hypolipidemic Properties of Probiotic-Enriched Fermented Camel Milk Combined with Salvia officinalis Leaves Hydroalcoholic Extract in Streptozotocin-Induced Diabetes in Rats. *Antioxidants* **2022**, *11*, 668. [CrossRef] [PubMed]
3. Bailey, C.J.; Tahrani, A.A.; Barnett, A.H. Future glucose-lowering drugs for type 2 diabetes. *Lancet Diabetes Endocrinol.* **2016**, *4*, 350–359. [CrossRef] [PubMed]
4. Beidokhti, M.N.; Jäger, A.K. Review of antidiabetic fruits, vegetables, beverages, oils and spices commonly consumed in the diet. *J. Ethnopharmacol.* **2017**, *201*, 26–41. [CrossRef] [PubMed]
5. Chouikhi, A.; Ktari, N.; Bardaa, S.; Hzami, A.; Ben Slima, S.; Trabelsi, I.; Ben Salah, R. A novel probiotic strain, Lactiplantibacillus plantarum LC38, isolated from Tunisian camel milk promoting wound healing in Wistar diabetic rats. *Arch. Microbiol.* **2022**, *204*, 1–13. [CrossRef] [PubMed]
6. Konuspayeva, G.; Faye, B. Recent advances in camel milk processing. *Animals* **2021**, *11*, 1045. [CrossRef] [PubMed]
7. El-Said, E.-S.E.-S.; El-Sayed, G.R.; Tantawy, E. Effect of camel milk on oxidative stresses in experimentally induced diabetic rabbits. *Vet. Res. Forum* **2010**, *1*, 30–43.
8. Zheng, Y.; Wu, F.; Zhang, M.; Fang, B.; Zhao, L.; Dong, L.; Ge, S. Hypoglycemic effect of camel milk powder in type 2 diabetic patients: A randomized, double-blind, placebo-controlled trial. *Food Sci. Nutr.* **2021**, *9*, 4461–4472. [CrossRef] [PubMed]
9. Kilari, B.P.; Mudgil, P.; Azimullah, S.; Bansal, N.; Ojha, S.; Maqsood, S. Effect of camel milk protein hydrolysates against hyperglycemia, hyperlipidemia, and associated oxidative stress in streptozotocin (STZ)-induced diabetic rats. *J. Dairy Sci.* **2021**, *104*, 1304–1317. [CrossRef] [PubMed]
10. Manaer, T.; Yu, L.; Nabi, X.-H.; Dilidaxi, D.; Liu, L.; Sailike, J. The beneficial effects of the composite probiotics from camel milk on glucose and lipid metabolism, liver and renal function and gut microbiota in db/db mice. *BMC Complement. Altern. Med.* **2021**, *21*, 127. [CrossRef] [PubMed]

11. Fallah, Z.; Feizi, A.; Hashemipour, M.; Kelishadi, R. Effect of fermented camel milk on glucose metabolism, insulin resistance, and inflammatory biomarkers of adolescents with metabolic syndrome: A double-blind, randomized, crossover trial. *J. Res. Med. Sci.* **2018**, *23*, 32. [CrossRef] [PubMed]
12. Xu, R.; Xiu, L.; Zhang, Y.; Du, R.; Wang, X. Probiotic and hepatoprotective activity of lactobacillus isolated from Mongolian camel milk products. *Benef. Microbes* **2019**, *10*, 699–710. [CrossRef] [PubMed]

biology and life sciences forum

MDPI

Proceeding Paper

Gastrointestinal Parasite Infestation of the Dromedary Camel (*Camelus dromedarius*) in Southern Algeria †

Ilyes Bouasla [1,*], Meriem Mekroud [1], Nadjet Amina Khelifi Touhami [1,2], Madjed Dib [3], Housna Bouhabila [1], Samia Daif [1], Nassim Ouchene [1,2], Amal Titi [1] and Ahmed Benakhla [4]

1 PADESCA Laboratory Research, Institute of Veterinary Sciences Khroub (ISVK), Mentouri Brothers University, Constantine 1, Constantine 25000, Algeria; mekroud.meriem@gmail.com (M.M.); khelifi.ouch.nadjet@gmail.com (N.A.K.T.); bouhabila.housna@gmail.com (H.B.); daif.samia@gmail.com (S.D.); ouchenassim@gmail.com (N.O.); titiamal88@gmail.com (A.T.)
2 Institute of Veterinary Sciences, University Saad Dahlab Blida 1, Street Soumaa, Blida 09000, Algeria
3 Camelin Research Unit, "Protection of Ecosystems in Arid and Semi-Arid Zones" Research Laboratory, Kasdi Merbah Ouargla University, Route de Ghardaia, BP. 511, Ourgla 30000, Algeria; dib.madjed@univ-ouargla.dz
4 Parasitology Laboratory, Chadli Bendjedid University, El Tarf 36000, Algeria; benakhlaahmed@gmail.com
* Correspondence: bouasla.ilyes.veto@gmail.com
† Presented at the 10th International Seminar of Veterinary Medicine: Camelids in Algeria & Maghreb, Constantine, Algeria, 20–21 December 2022.

Abstract: This study aims to evaluate the prevalence and identification of gastrointestinal parasites in fecal samples of dromedary camels (*Camelus dromedarius*) in Algeria based on microscopic examination. A total of 46 fresh fecal samples were collected in the southern Algerian towns of Adrar, Tindouf, Ourgla, and Ain Salah, and nine samples from camel farms in Constantine. Samples were examined with the flotation technique. Results showed an infestation rate of dromedaries of 32.6% (15/46), with seven different gastrointestinal species: 25.4% protozoa (*Balantidium coli*, *Eimeria dromederi*), 16.3% nematodes (*Nematodirus* spp., *Trichuris* spp., others strongles), and 3.6% cestodes (*Moniezia* spp.). Dromedaries are more infested with protozoa than with nematodes or cestodes ($p < 0.05$). The type of sex has no significant influence on the rate of parasitic infestation. The Sahraoui breed (70%) appears more infested compared to the Tergui breed (31.42%) ($p < 0.05$). Dromedaries originating from the Ouargla region (73.68%) are the most infested compared to those originating from Adrar (18.18%), Ain Salah (10%), and Tindouf (33.33%) ($p < 0.05$).

Keywords: dromedary camels; gastrointestinal parasites; southern Algeria; flotation technique

check for updates

Citation: Bouasla, I.; Mekroud, M.; Khelifi Touhami, N.A.; Dib, M.; Bouhabila, H.; Daif, S.; Ouchene, N.; Titi, A.; Benakhla, A. Gastrointestinal Parasite Infestation of the Dromedary Camel (*Camelus dromedarius*) in Southern Algeria. *Biol. Life Sci. Forum* **2023**, 22, 19. https://doi.org/10.3390/blsf2023022019

Academic Editors: Amira Leila Dib, Said Boukhechem, Hithem Bougherara and El-Hacene Bererhi

Published: 12 April 2023

1. Introduction

The camel, commonly called the ship of the desert in Arabic, is an important animal. They live in desert areas due to their ability to withstand very harsh conditions (high temperature and drought), to provide milk and meat, to be used as a means of transport [1,2], and to digest poor forage compared to other domestic ruminants [3]. In Algeria, the camel farming sector (354,465 camel heads) has a substantial contribution to make toward covering the growing gap in protein and dairy products. Several studies have reported the occurrence of different gastrointestinal parasites in camels in different parts of the world [4,5]. However, few studies have examined these diseases in Algeria. This animal is frequently infested by gastrointestinal parasites, which decreases productivity [6,7]. Among the many pathologies caused by these parasites, helminthiasis represents an important internal parasitosis affecting camels. The aim of this study is to evaluate the infestation by gastrointestinal parasites in local and transhumant dromedary camels in southern Algeria (Tindouf, Adrar, Ouergla, and Ain Salah) and Constantine.

2. Materials and Methods

Nine faecal samples of transhumant dromedaries from the Neili population were collected in Constantine in 2019 and treated with the flotation technique before being examined under a microscope. Forty-six faecal samples from the local Tergui and Sahraoui camel populations were collected in March 2022 in four regions of southern Algeria (Ourgla, Adrar, Tindouf, and Ain Salah), stored in cold storage, and transported to the laboratory. The samples were treated using the flotation technique and examined under a microscope. Each fecal sample was examined by direct smear and the simple flotation method using a saturated salt solution [8]. Age, sex, and population information were collected.

An ANOVA test was used for statistical analysis. The difference was considered significant at $p < 0.05$.

3. Results and Discussion

Seven different gastrointestinal parasite species were identified in camels: four nematodes, one cestode, and two protozoa. Dromedaries are more infested with protozoa than with nematodes or cestodes ($p < 0.05$). The type of sex has no significant influence on the rate of parasitic infestation. The Sahraoui breed (70%) appears more infested compared to the Tergui breed (31.42%) ($p < 0.05$). Dromedaries originating from the Ouargla region (73.68%) are the most infested compared to those originating from Adrar (18.18%), Ain Salah (10%), and Tindouf (33.33%) ($p < 0.05$).

The prevalence rate of gastrointestinal parasites found in our study, *Eimeria dromaderi*, is present in all southern regions in local camels with a prevalence rate of 10.5–45%, which is higher than 6.7% [9] and 5.7% [10], and we can see that the animals of the Ouergla region are the most infested compared with other regions.

Balantidium coli was recorded in Ouergla in three samples (15.7%), higher than the 6.7% found in Egypt by Ref. [9].

The prevalence of *Nématodirus* spp. was 55% in the transhumant camels of Constantine and 10% in local camels in Ouergla, compared with results obtained by Ref. [11] in Algeria with a prevalence of 23%. *Strongyloides* spp. were detected in one camel sample (5.2%), almost similar to the(4%) obtained by Ref. [11] in Algeria. *Trichuris* spp. prevalence of 10% was higher than the 1% and 2.23% obtained by Refs. [11] and [10], respectively. A cestode infestation (*Moniezia* egg) was observed in two camels examined with a rate 10% higher than the 7.5% obtained in Egypt by Ref. [9]. The results obtained with the flotation technique show a low infestation with protozoa (*Eimeria dromaderi* and *Balantidium coli*) and a weak infestation with *Trichuris* spp., which was observed only in Ourgla.

4. Conclusions

The camel gastrointestinal parasites identified in this study show similarities to other ruminant infestations by various species of helminths and protozoa. The gastrointestinal parasites of camels identified in this study testify that there is similarity with the infestation of other ruminants by different species of helminths and protozoa. Other interesting lines of research can focus on the life cycle and the economic impact of parasites.

Author Contributions: Conceptualization I.B., A.T., M.D., M.M., S.D. and H.B.; methodology, I.B., M.D., A.T., N.O. and N.A.K.T.; validation, I.B., A.B. and A.T.; investigation, I.B. and A.T.; writing—original draft preparation I.B., M.M., A.T. and N.A.K.T.; writing—review and editing, I.B. and M.M.; supervision, A.B. and A.T. All authors have read and agreed to the published version of the manuscript.

Funding: This research received no external funding.

Institutional Review Board Statement: Not applicable.

Informed Consent Statement: Not applicable.

Conflicts of Interest: The authors declare no conflict of interest.

References

1. Faye, B.; Jaouad, K.M.; Bhrawi, A.; Senoussi, M.; Bengoumi, M. Elevage camelin en Afrique du Nord: état des lieux et perspectives. *Rev. Elev. Med. Vet. Pays Trop.* **2014**, *67*, 213–221. [CrossRef]
2. Đuričić, D.; Kilvain, I.; Samardžija, M. Fiziologija rasplođivanja kamelida-Anatomija spolnih organa i spolna zrelost I. *dio. Vet. Stanica* **2020**, *51*, 353–362. (In Croatian) [CrossRef]
3. Kayouli, C.; Jouany, J.P.; Dardillat, C.; Tisserand, J.-L. Particularités physiologiques du dromadaire: Conséquences pour son alimentation. *Options Méditerr.* **1995**, *13*, 143–155. (In French)
4. Sharif, L.; Al-Qudah, K.M.; Al-Ani, F. Prevalence of gastro-intestinal parasites in one-humped camels (Camelus dromedarius) in Jordan. *J. Camel Pract. Res.* **1997**, *4*, 67–69.
5. Magzoub, M.; Omer, O.H.; Haroun, E.M.; Mahmound, O.M. Effect of season on gastrointestinal nematode infection in Saudi Arabian camels (Camelus dromedarius). *J. Camel Pract. Res.* **2000**, *7*, 107–108.
6. Richard, D. Haemonchosis in camels. *Rev. Elev. Med. Vet. Pays Trop.* **1989**, *42*, 45–53. (In French) [PubMed]
7. Mahmuda, A.; Mohammed, A.A.; Alayande, M.O.; Habila, Y.I.; Lawal, M.D.; Usman, M.; Raji, A.; Yahaya, M.S.; Suleiman, N. Prevalence and distribution of gastrointestinal parasites of working camels in Sokoto metropolis. *Vet. World* **2014**, *7*, 108–112. [CrossRef]
8. Cebra, C.K.; Stang, B.V. Comparison of methods to detect gastrointestinal parasites in llamas and alpacas. *J. Am. Vet. Med. Assoc.* **2008**, *232*, 733–741. [CrossRef] [PubMed]
9. El-Khabaz, K.A.; Abdel-Hakeem, S.S.; Arfa, M.I. Protozoan and helminthes parasites endorsed by imported camels (*Camel dromedaries*) to Egypt. *J. Parasit. Dis.* **2019**, *43*, 607–615. [CrossRef] [PubMed]
10. Bouragba, M.; Laatamna, A.K.; Cheddad, F.E.; Baroudi, D.; Houali, K.; Hakem, A. Gastrointestinal parasites of dromedary camel (*Camelus dromedarius*) in Algeria. *Vet. World* **2020**, *13*, 1635–1640. [CrossRef] [PubMed]
11. Saidi, R.; Mimoune, N.; Chaibi, R.; Abdelouahed, K.; Khelef, D.; Kaidi, R. Camel gastrointestinal parasites in southern Algeria. *Vet. Stanica* **2022**, *53*, 283–294. [CrossRef]

biology and life sciences forum

MDPI

Proceeding Paper

Reproductive Troubles: Cases Report in Camels—First Caesarean in Camilidae in Algeria †

Said Fettata

Independent Researcher, Metlili El Jadida 47000, Algeria; fettatasaid@gmail.com

† Presented at the 10th International Seminar of Veterinary Medicine: Camelids in Algeria & Maghreb, Constantine, Algeria, 20–21 December 2022.

Abstract: The camel is a highly valuable animal which contributes effectively to the welfare of people in difficult environments such as the Algerian Sahara. Reproduction in the camel is not as well understood as in more common species of domestic animals. Reproductive problems in the camel are not researched as, for example, in the bovine and small ruminants. Our study demonstrates a clinical case in the Ghardaïa district: cesarean section in female's camels. The incidence of camel dystocia does not differ from that of bovines. The etiologies of dystocia include uterine torsion, carpal flexion, lateral deviation of the head and hock and hip flexion of the fetus. However, the camel fetus survives dystocia better than the equine fetus, and the camel is a good subject for cesarean section. Cesarean section could be performed on the left flank using xylazine sedation and local regional or infiltration anesthesia. A camel, 17 h in dystocia, delivered a live fetus via cesarean section. The camel placenta is diffuse epitheliochorial type, and placental retention subsequent to parturition is rare. The camel placenta is expelled within 49 min to 6 h of calving.

Keywords: dromedary camel; reproductive troubles; cesarean section; Ghardaïa

check for updates

Citation: Fettata, S. Reproductive Troubles: Cases Report in Camels—First Caesarean in Camilidae in Algeria. *Biol. Life Sci. Forum* **2023**, *22*, 20. https://doi.org/10.3390/blsf2023022020

Academic Editors: Amira Leila Dib, Said Boukhechem, Hithem Bougherara and El-Hacene Bererhi

Published: 13 April 2023

1. Introduction

The first veterinary cesarean section on a cow was performed by Morange, a veterinary practitioner, in Lestern, in 1813 [1]. Cesarean section is a surgical procedure which requires good anatomical, physiological, propaedeutic as well as therapeutic knowledge.

It is the most common surgical operation in rural veterinary practice [2]. Cesarean section in mammals begins by standing or lying while restrained. It offers a choice of operating sites with shaving and disinfection, premedication (local or loco-regional anesthesia), laparotomy, hysterotomy, exteriorization of the newborn, uterine suture, abdominal wall suture and treatment post-surgery [2].

Cesarean sections on camels are less documented than cesarean sections on cows, mares, and small ruminants, which makes the operation the least common in rural veterinary practice and leaves a negative impact on the survival of dystocic camels.

The natural process of parturition in dromedary camels (Camelus dromedarius) can span over a variable period of time ranging from 2 to 6 h [3].

There is little clinical information regarding cesarean section in *C. dromedarus*. Therefore, the aims of this study were:

1—To compare the cesarean section in camels and bovines.

2—To present such techniques and facilitate the act.

3—To elaborate a guide to cesarean sections for veterinary practitioners in Algeria.

2. Materials and Methods

2.1. Pre-Operative Evaluation

The subject was a 15-year-old female dromedary camel (Sahraoui breed). General examination: the female presented a weakness and was treated with fluid therapy. Obstetric examination revealed an irreducible 180° post cervical torsion, right-hand direction.

2.2. Surgical Instruments

Shaving equipment, disinfectant solution, a scalpel and two blades, Lister scissors, curved scissors with a foam tip and straight, a needle holder, one curved needle with a round section to sew the uterus, two S-shaped needles with triangular section for muscular and skin sutures, absorbable thread for internal and muscular sutures, non-absorbable thread for skin sutures, traumatic forceps with rubber jaws for gripping the uterus, claw pliers for gripping the peritoneum, right hemostatic forceps, compresses, 20 mL syringes for anesthesia products, local anesthetic products (lidocaine), surgical gloves, sterile drapes and drapes, high-quality ropes for restraint [4].

The objective of the intervention was to compare a cesarean in a camel and a cow.

3. Results and Discussion

(1) Restraint: Good compression in sternal recumbent position (squatting animal).
(2) Place of election: Hollow of the left flank.
(3) Preparation of the place of election: The place of election must be prepared to avoid the risks of subsequent infection to the greatest extent possible. The intervention site is mowed and shaved. The region is largely soaped, then disinfected, and dried, then covered with a sterile operating drape.
(4) Anesthesia: It is usually local anesthesia administered by infiltration (lidocaine).
(5) Operating technique:
    (a) Laparotomy: Skin incision over a length of 30 to 40 cm (Figure 1a). Then, the muscular skin, then the abdominal tunic, the fused fascia of the obliques, the right of the abdomen, the fascia of the transverse of the abdomen and the peritoneum.
    (b) Hysterotomy: Externalization of the uterus. Incision of the uterus at the level of the great curvature, over a length of about 30 cm.
    (c) Extraction of the camel (Figure 1b): We start with the extraction of the forelegs, then the head with the long neck, then the rest of the body (at this time, the camel is alive (Figure 1c)).
    (d) Placental delivery: This is performed manually, followed by local antibiotic therapy.
    (e) Suture of the uterus: first suture: Utrecht suture, second suture (optional): Cushing suture.
    (f) External washing of the uterus and dusting of the wound with penicillin.
    (g) The uterus is returned to the abdominal cavity after careful checking.
    (h) Intra-abdominal antibiotic therapy.
    (i) Parietal sutures. Peritoneum, muscular (single overlock), skin (in separate "U" shape).
(6) Post-operative care: General antibiotic therapy for 8 days, non-steroidal anti-inflammatory drug for 5 days, followed by cleaning and disinfection of the skin wound.

The method of restraint in camels requires experts' involvement because of the high level of difficulty. It is very hard to carry out the operation on a standing camel. The indications for cesarean in camels are the same as in cows, and the surgical procedure of the cesarean section in the camel is the same as in the cow. The extraction of the uterus in the camel is very easy compared to that of the cow. The extraction of the newborn camel is a little difficult given the length of the neck and legs. The placenta in camels is of diffuse type and is not cotyledonary, as it is in bovines. On this occasion, the cesarean section ended with the camel living and in good health. The involution of the uterus after placental delivery in the camel is very rapid; at the end of the last sutures, the uterus reaches a very reduced size, which explains the return of ovarian cyclicity a few days to a few weeks after giving birth. On expulsion, the young dromedary is surrounded by a thin membrane called the "epidermal" membrane, which ruptures in response to the first movements. This membrane seems to play a thermoregulatory role and opposes the drying out of the organism. Unlike other species, the dromedary female does not lick her infant, but only sniffs it. The offspring are recognized mainly by sniffing. All camel dystocia without the intervention of veterinary obstetrics ends in slaughter or death. After three weeks, the camel joins the herd in the middle of the desert, to the satisfaction of the owner.

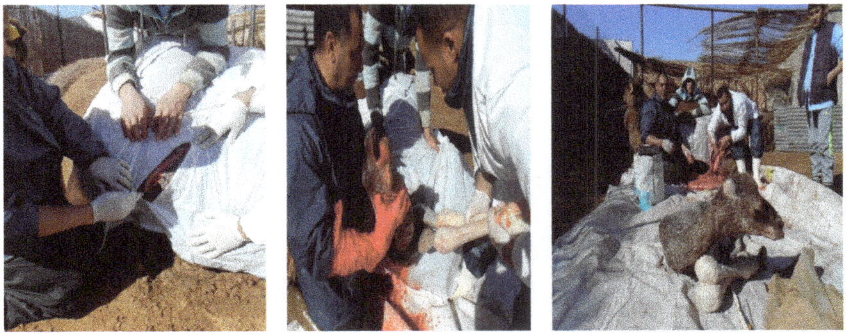

(a) (b) (c)

Figure 1. Cesarean section in camels. This was a great success. (**a**) Skin incision. (**b**) Extraction of the camel. (**c**) The camel is alive.

4. Conclusions

Camel obstetrics is very similar to that of other species, but remains unknown by the majority of Algerian veterinarians. Early surgical intervention is significantly associated with better survival rates of both the dam and calf. The application of the same cesarean protocol in cows and in camels resulted in a motivating success and contributed to the amplification of this act, which will prevent the loss of the camel herd. This first cesarean in Algeria left a positive impact on breeders, and the number of cases of interventions will increase each year.

Funding: This research received no external funding.

Institutional Review Board Statement: Not applicable.

Informed Consent Statement: Not applicable.

Data Availability Statement: Not applicable.

Conflicts of Interest: The author declares no conflict of interest.

References

1. Titinger, P. Historique de L'opération Césarienne: Étude Comparative de L'évolution et Obstétriques Humaine et Vétérinaire. Ph.D. Thesis, Université Claude Bernard, Lyon, France, 1996. (In French)
2. Hanzen, C.; Lourtie, O.; Ectors, F. La césarienne dans l'espèce bovine. *Ann. Méd. Vét.* **1999**, *143*, 65–90. (In French)
3. Roberts, S. *Veterinary Obstetrics and Genital Diseases*; Literary Licensing, LLC: Whitefish, MT, USA, 2012.
4. Cardot, T. Prise en Charge de la Douleur et de L'inflammation Postopératoires Lors de Césarienne Chez la Génisse Allaitante et Évaluation de Paramètres de Reproduction. Ph.D. Thesis, Alfort National Veterinary School, Maisons-Alfort, France, 2018. (In French)

biology and life sciences forum

MDPI

Proceeding Paper

Histological Variations in the Uterine Mucosa during the Postpartum Period in Camels (*Camelus dromedarius*) †

Rabah Kelanemer [1,*], Djallel Adel [1], Hocine Ziam [1,2], Bachir Medrouh [3], Amina Saidi [1], Yasmine Rahmoune [4], Naima Dellal [4] and Said Fettata [5]

1 Institute of Veterinary Sciences, University Saad Dahleb Blida 1, Blida 09000, Algeria; adel_djallel@yahoo.fr (D.A.); veziamocine@gmail.com (H.Z.); minasaidi@gmail.com (A.S.)
2 Laboratory of Biotechnology, Environment and Health, University Saad Dahleb Blida 1, Blida 09015, Algeria
3 The Agropastoralism Research Center (CRAPast), Djelfa 17000, Algeria; medrouhbaepidemio15@gmail.com
4 High Commission for the Development of the Steppe, Djalfa 17000, Algeria; rahmoune.vet@gmail.com (Y.R.); naimadel16@gmail.com (N.D.)
5 Independent Researcher, Metlili El Jadida 47000, Algeria; fettatasaid@gmail.com
* Correspondence: veto_toxico@yahoo.fr
† Presented at the 10th International Seminar of Veterinary Medicine: Camelids in Algeria & Maghreb, Constantine, Algeria, 20–21 December 2022.

Abstract: This work aims to determine the main histological changes in the endometrium of female camels during the postpartum period (recovered from the uterine epithelium). For this, successive samples of uterine mucosa were taken from the left uterine horn of females from the 3rd, 5th, 7th, 11th, 15th, 18th, and 21st postpartum day. The samples of the uterine biopsies were carried out on 10 camels. In this study, it appears that the recovery of the epithelium of the uterine endometrium is short (3 weeks) and comparable to that of mares. In conclusion, this study verified the hypothesis of short uterine involution in camels and the rapid resumption of ovarian activity.

Keywords: camel uterine mucosa; histology; postpartum

check for updates

Citation: Kelanemer, R.; Adel, D.; Ziam, H.; Medrouh, B.; Saidi, A.; Rahmoune, Y.; Dellal, N.; Fettata, S. Histological Variations in the Uterine Mucosa during the Postpartum Period in Camels (*Camelus dromedarius*). *Biol. Life Sci. Forum* **2023**, 22, 21. https://doi.org/10.3390/blsf2023022021

Academic Editors: Amira Leila Dib, Said Boukhechem, Hithem Bougherara and El-Hacene Bererhi

Published: 13 April 2023

1. Introduction

Camel has several physiological particularities including those related to the reproduction [1]. It is a polyestrous species, with provoked seasonal with ovulation (ovarian cycle of the follicular type). Puberty and the start of reproduction begin late in life (from 3 to 4 years), with a long gestation period (from 12 to 13 months) [2]. These caracteristics lead to a low level of reproduction [3]. For local authorities, camel breeding constitutes an essential element for the development of arid and semi-arid zones. Our study constitutes a contribution to the understanding of the postpartum period, in particular, postpartum uterine involution. Some authors have reported that involution is short and the resumption of ovarian activity begins sooner in this species that is comparable to that of mares (from 15 to 20 days) [4]. However, the authors of Ref. [5] reported a long duration that is comparable to that of the cow (35 to 45 days). The search for a method of monitoring ovarian function, in particular, during the postpartum period, is necessary in order to increase reproductive performance. The main objective is verifying the hypothesis of histologically reduced uterine involution. Histological examination during a biopsy is a reliable technique for the evaluation of endometrial changes [6].

This review presents the easiest and most appropriate way to diagnose inflammatory or degenerative changes. It can also determine the moment of the end of uterine involution, complete endometrial restoration, and even the timing of the cycle. The objective is to determine the main histological changes in the endometrium in female dromedaries during the postpartum period; in light of the histological results of the endometrium, it appears that uterine involution in camels completes on the 21st postpartum day. The process of uterine involution occurs through the association of endometrial regeneration phenomena and vascular phenomena.

2. Materials and Methods

An introduction and the methods of the biopsy are included here. Histological examination during a biopsy is a reliable technique for the evaluation of endometrial changes [6]. This review presents the simplest and most appropriate way to diagnose inflammatory or degenerative changes. It can also determine the end of uterine involution, complete endometrial restoration, and even the timing of the cycle. The objective is to determine the main histological changes in the endometrium in female camels during the postpartum period. By successive histological sections of uterine mucosa were taken from females from the 3rd postpartum day.

3. Results and Discussion

3.1. From 3rd Postpartum Day

Our results show that on the third day after parturition the epithelium of the uterine endometrium presents the following characteristics: Discontinuous, desquamated surface epithelium that is detached in places (Figure 1a). On Figure 1b we can observe a simple prismatic surface epithelium with cells with large nuclei.

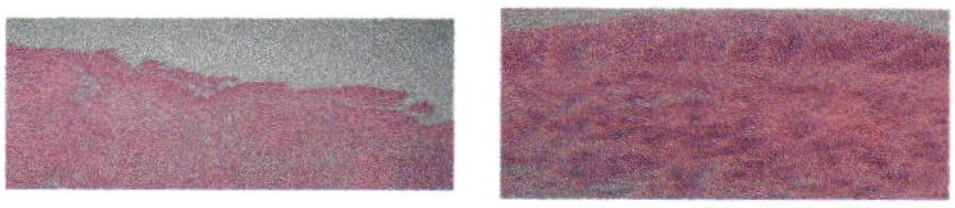

(**a**) (**b**)

Figure 1. (**a**) Histological section of the endometrium on the third day postpartum (10). (**b**) Histological section of the endometrium on the third day postpartum (40).

3.2. Between 5th and the 7th Day Postpartum

During the period between the fifth and the seventh day of the postpartum one observes a epithelium is more intact than that of the surface, and it is darker. The chorion is loose and characterized separately by less important vascularization. The number of uterine glands is a few or they are absent (Figure 2). They have small circular sizes and are scattered in the chorion.

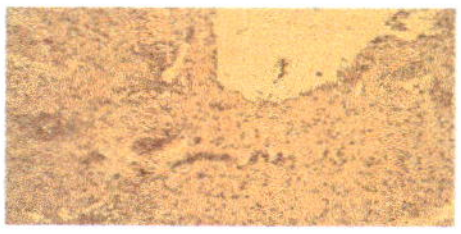

Figure 2. Histological section of the endometrium between the fifth and seventh day postpartum (10).

3.3. From 10th Postpartum Day

From the tenth day after parturition there is appearance of the uterine glands. The uterine glands are more numerous compared to the number of those at the previous stage. These glands are small in size, disseminated in the chorion, and their lumen can be expanded or reduced (Figure 3).

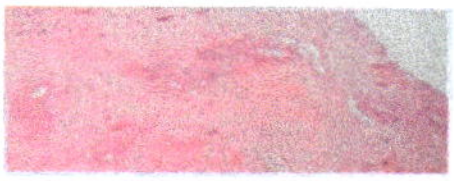

Figure 3. Histological section of the endometrium on the tenth day postpartum (40).

3.4. From 11th to 14th Postpartum Day

During the second week we observe: The surface epithelium is poorly regenerated and desquamated in a few places (Figure 4a). The uterine glands underwent an increase in number and size, they are circular or elongated in shape, they can be scattered in the lamina propria or organized in clusters, and they are surrounded by blood vessels (Figure 4b).

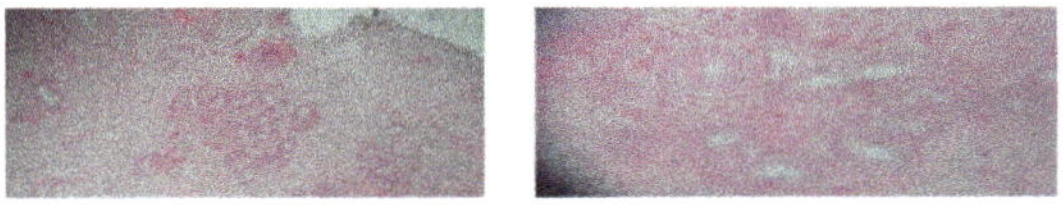

(a) (b)

Figure 4. (**a**) Histological section of the endometrium between the eleventh and fourteenth day postpartum (10). (**b**) Histological section of the endometrium between the eleventh and fourteenth day postpartum (40).

3.5. From 15th to 18th Postpartum Day

At the beginning of the second week the surface epithelium is not yet completely restored, and it is of the simple cylindrical type. Many uterine glands are disseminated in the chorion. Neovascularization is very important as is increases the number and size of glands (Figure 5).

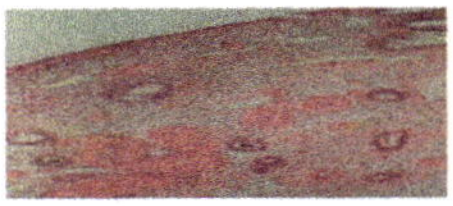

Figure 5. Histological section of the endometrium between the fifteenth and eighteenth day postpartum (40).

3.6. From 19th to 21st Postpartum Day

At the end of the third week at magnification (10) (Figure 6a), the surface epithelium appears to be more intact, uniform, and continuous over its entire surface.

At the highest magnification (40) (Figure 6b), the epithelium appears to be simple and cylindrical. In light of the histological results of the endometrium, it appears that uterine involution in camels completes on the 21st postpartum day. These results agree with those in Ref. [7], which reports a duration of 40 days. We also used the method of transrectal palpation. Our results are in agreement with those declared by the authors of Refs. [4,8], who reported a duration from 15 to 28 days. The uterine lining is completely restored from the 18th postpartum day.

The process of uterine involution occurs through the association of endometrial regeneration phenomena and vascular phenomena.

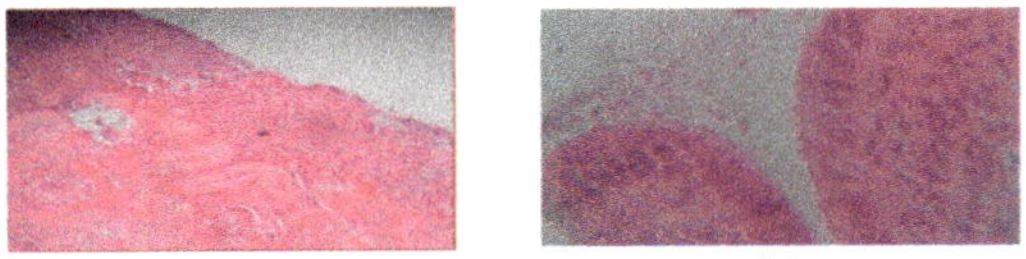

(a) (b)

Figure 6. (**a**) Histological section of the endometrium between the nineteenth and twenty-first days postpartum(10). (**b**) Histological section of the endometrium between the nineteenth and twenty-first days postpartum (40).

4. Conclusions

It appears, in this brief and clear study that the restoration of the epithelium of the uterine endometrium completes at the end of the third week.

Author Contributions: Conceptualization, R.K., H.Z. and D.A.; methodology, B.M., A.S. and Y.R.; validation, N.D. and S.F.; writing—original draft preparation, R.K.; writing—review and editing, H.Z. and D.A.; project administration and interpretation of data, R.K.; the decision to publish the results, R.K., H.Z., A.S. and D.A. All authors have read and agreed to the published version of the manuscript.

Funding: This research received no external funding.

Institutional Review Board Statement: During this study, no invasive procedure was performed during the sampling.

Informed Consent Statement: Not applicable.

Data Availability Statement: Not applicable.

Acknowledgments: Our thanks go to the director of the experimental station at the university veterinary institute of Blida, the director of the high commission for the development of the steppe, and the director of the research center on agropastoralism for their help.

Conflicts of Interest: The authors declare no conflict of interest.

References

1. Zarrouk, A.; Souilem, O.; Beckers, J. Actualités sur la reproduction chez la femelle dromadaire (*Camelus dromedarius*). *Rev. Elev. Med. Vet. Pays Trop.* **2003**, *56*, 95–102. (In French) [CrossRef]
2. Skidmore, J.A. Reproductive physiology in female old world camelids. *Anim. Reprod. Sci.* **2011**, *124*, 148154. [CrossRef] [PubMed]
3. Kelanemer, R.; Antoine-Moussiaux, N.; Moula, N.; Abu-Median, A.A.K.; Hanzen, C.H.; Kaidi, R. Effect of nutrition on reproductive performance during the peri-partum period of female camel (*Camelus dromaderius*) in Algeria. *J. Anim. Vet. Adv.* **2015**, *14*, 192–196.
4. Vaughan, J.; Tibary, A. Reproduction in female South American camelids: A review and clinical observations. *Small Rumin. Res.* **2006**, *61*, 259–281. [CrossRef]
5. Musa, B.; Makawi, S. Involution of the uterus and the first postpartum heat in the camel (*Camelus dromedarius*). In Proceedings of the Conference on Animal Production in Arid Zones, Damascus, Syria, 7–12 September 1985.
6. Tibary, A.; Anouassi, A. Uterine infections in camelidae. *Vet. Sci. Tomorrow* **2001**, *1*, 1–12.
7. El-Wishy, A. Functional morphology of the ovaries of the dromedary camel. In Proceedings of the First International Camel Conference, Dubai, United Arab Emirates, 2–6 February 1992.
8. Derar, R.; Ali, A.; Al-Sobayil, F. The postpartum period in dromedary camels: Uterine involution, ovarian activity, hormonal changes, and response to GnRH treatment. *Anim. Reprod. Sci.* **2014**, *151*, 186–193. [CrossRef] [PubMed]

biology and life sciences forum

MDPI

Proceeding Paper

Comparison of the CMT Test Results Carried Out on Camels and Bovines †

Brahim Bouchoucha *, Noureddine Zeghilet and Omar Bouaziz

GSPA Research Laboratory (Management of Animal Health and Productions), Institute of Veterinary Sciences of El-Khroub, Université Frères Mentouri Constantine 1, Constantine 25000, Algeria; nzeghilet@yahoo.fr (N.Z.)

* Correspondence: brahimbouchocuha4@gmail.com or ibrahim.bouchoucha@doc.umc.edu.dz

† Presented at the 10th International Seminar of Veterinary Medicine: Camelids in Algeria & Maghreb, Constantine, Algeria, 20–21 December 2022.

Abstract: Subclinical intra-mammary infection is a very current disease in dairy females; several early tests exist for their detection: California Mastitis Test (CMT), Electrical Conductimetry of milk (CE), pH indicator papers. In this study, the milk of 104 cows (416 quarters) was tested using CMT (Raidex®) to first estimate the prevalence of the disease and to identify the effect of age, stage of lactation and position of neighborhoods on its prevalence; second, we looked for data related to camels from another study carried out on 57 camels (in Algiers (ENVH), starting with the hypothesis that the camel is more resistant. The prevalence rate (CMT > 1) is around 45% for cows and 15% for quarters. The study conducted in Algiers yielded a rate of 67% (for camels and 35% for quarters, i.e., an increase of 130% for quarters and 50% for animals. Our study emphasizes the impact of the last stage of lactation compared to the first (35% vs. 15%), which is in agreement with the other study; on the other hand, the effect of the posterior position of the teats and age was contradictory. The differences between the rates were not significant at $p < 0.05$ between neither the animals or quarters (Chi 2: 53.24 and 71.41).

Keywords: subclinical mastitis; comparison; prevalence; cattle; camels

check for updates

Citation: Bouchoucha, B.; Zeghilet, N.; Bouaziz, O. Comparison of the CMT Test Results Carried Out on Camels and Bovines. *Biol. Life Sci. Forum* **2023**, *22*, 16. https://doi.org/10.3390/blsf2023022016

Academic Editors: Amira Leila Dib, Said Boukhechem, Hithem Bougherara and El-Hacene Bererhi

Published: 10 April 2023

1. Introduction

Mastitis is considered the most frequent and harmful pathology [1,2]; its subclinical form causes many biological changes in milk, such as affecting the somatic cell content [3], salts (Na Cl) content [4] and the acidity of the milk [5], which requires indirect screening tests, such as the California Mastitis Test(CMT) [5], Electrical Conductivity (EC) of milk and pH indicator papers [6]. These tests are frequently used by researchers and some veterinarians who testify to the existence of inflammation in udders. The results of this method make it possible to obtain the rate of prevalence and study the impact of different risk factors on this rate to help in prevention. Our objective in this study was to study the prevalence of this disease and compare the results obtained from dairy cattle with those obtained from camels. In the end, our objective was to study the effect of age, teat position and the month of lactation on the prevalence rate.

2. Material & Methods

2.1. Farms and Animals

For cows, samples were taken from regions of Mila, Guelma and Constantine. For camels [6], samples were taken from farms in Biskra and Ouargla. The distribution of animals according to age and the month of lactation is as follows: For camels, we have 40 primipareous, 12 in middle age and 5 camels over 17 years old. For the stage of lactation, there were 14 camels in the beginning and, at the end of lactation, 43 camels. For cattle, according to age, there were 18 primiparous, 79 in middle age and 7 old cows; for the stage of lactation, 74 were in the early stage of lactation and 30 cattle were at the end of lactation.

2.2. *CMT*

A test was applied. Its principle was based on the use of Teepol, which leads to cells bursting and the precipitation of their DNA; the technique is based on the mixture of milk (2 mL) and 2 mL of reagent in cups on an appropriate test plate, which is then stirred and read according to the degree of coagulation. In terms of reading, gelation corresponds to a positive test. The absence of gelation is negative; if there is a gel that disappears after a few seconds, it is considered suspicious.

2.3. *Prevalence*

The prevalence value is calculated in the concordance with the equation.
Prevalence = incidence × duration [5].

2.4. *Statistical Analysis*

For statistical analysis, we have used Excel software and the Chi 2 test, with Excel Logical for calculating meaning and drawing graphics and the Chi 2 test for comparing between different meanings.

3. Results & Discussion

In Table 1, we can find results of different prevalences of disease in function of stage of lactation, age and udders position.

Table 1. Results of study on camels and cows.

	Stage of Lactation		**Age**			**Udders**	
	Beginning	End	Primiparous	Middle	Old	Post	Ant
Camelins (%) [6]	14 (25)	43 (75)	40 (70)	12 (21)	5 (9)	71 (33)	80 (37)
Bovins (%)	74 (71)	30 (29)	18 (17)	79 (76)	7 (17)	34 (16)	13 (6)

Post: Posterior; Ant: Anterior.

The prevalence rate (CMT > 1) is around 45% (47/104) for the cattle studied and 15% (62/416) for their quarters. In addition, the study carried out in camels yielded prevalence rates of 67% (38/57) for camels and 35% (76/216) for teats, respectively, i.e., an increase of 130% for teats and 50% for animals; Ref. [5] reports that camels are as susceptible to mastitis as other species. The differences between rates were not significant at $p < 0.05$, neither for animals nor quarters (Chi 2= 53.24 and 71.41) (Figure 1).

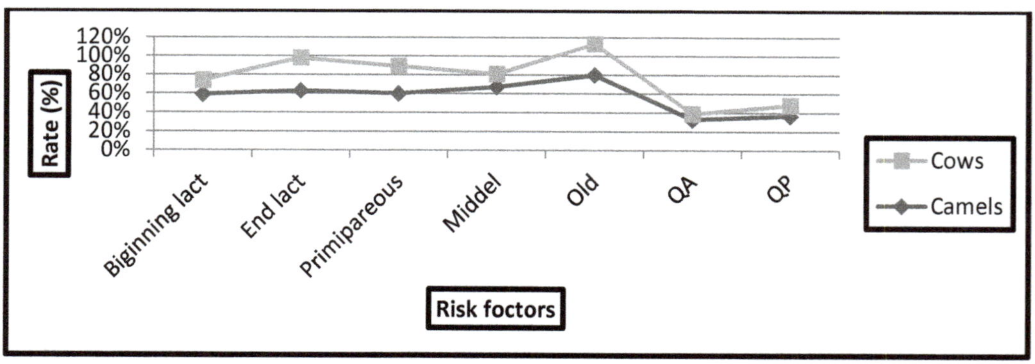

Figure 1. Effects of different risk factors on the rate of subclinical mastitis according to the species. lact = lactation; AQ = Anteriors Quarters; PQ = Posteriors Quarters.

Our study emphasizes the impact of the last stage of lactation compared to the first (35% vs. 15%), which is in agreement with the other study (63% vs. 59%); on the other hand [7,8] evoke the effect of the first months. The effect of the posterior position of the cattle's teats compared to the anterior (11% vs. 6%) is in agreement with [9] but is in opposition to the study on camels (33% for hind vs. 37% for fronts). Regarding the effect of age, as mentioned by the study on camels (60%, 67% and 80%), depended on the classification of the age, even for [9], in opposition to our study, wherein we obtained fluctuating frequencies (29%, 14% and 33%).

4. Conclusions

This study confirms that Subclinical Mastitis is an existing disease in our country. The disease affects bovines and camels with varying rates and is impacted by different risk factors (age, teats position and stage of lactation). Furthermore, it requires early detection methods for screening and further study of the risk factors for prevention.

Author Contributions: Conceptualization, B.B., N.Z. and O.B.; methodology, O.B.; software, B.B.; validation, B.B., N.Z. and O.B.; formal analysis, O.B.; investigation, B.B.; resources, B.B.; data curation, B.B.; writing—original draft preparation, B.B.; writing—review and editing, B.B.; visualization, N.Z.; supervision; B.B.; project administration, O.B.; funding acquisition, O.B. All authors have read and agreed to the published version of the manuscript.

Funding: This research received no external funding.

Institutional Review Board Statement: Not applicable.

Informed Consent Statement: Not applicable.

Data Availability Statement: The works consulted are detailed in the references.

Acknowledgments: The authors acknowledge the support and help from Khlifa Zineb veterinary doctor who give us his results of camels study and Many thanks to the breeders who collaborated, to the staff of the Hiba analysis laboratory at El-Khroub in Constantine and the Staff of the microbiology laboratory of the Institute of Veterinary Sciences at El-Khroub. Constantine, Algeria.

Conflicts of Interest: The authors declare no conflict of interest.

References

1. Durel, L.; Faroult, B.; Lepoutre, D.; Brouillet, P.; Le Page, P. Cattle mastitis (clinical and subclinical). Diagnostic and therapeutic approaches. Technical supplement. *Dépêche Vétérinaire* **2004**, *87*, 42. (In French)
2. Hanzen, C. *Propaedeutics of the Mammary Gland, Semiology and Individual Diagnosis of the Herd*; Faculty of Veterinary Medicine, University of Liege: Liège, Belgium, 2009; pp. 1–28. (In French)
3. Harmon, R.J. Physiology of mastitis and factors affecting somatic cell counts. *J. Dairy Sci.* **1994**, *77*, 2103–2112. [CrossRef] [PubMed]
4. Guerin, P. Staphylococcal mastitis in cows: Epidemiological aspects. In Proceedings of the Staphylococci and Public Health, Ninth GTV Rhône-Alpes Meetings, National Veterinary School of Lyon, Lyon, France, 18 June 1998; pp. 1–21. (In French)
5. Kumar, P.; Deora, A.; Himanshu, S.; Sharma, S.; Mittal, D.; Bhanot, V.; Prakash, A.; Yadav, R.; Diwaka, R.P. Bovine Mastitis: A Review. *Middle-East J. Sci. Res.* **2020**, *28*, 497–507.
6. Khelifa, Z. *Diagnosis of Subclinical Mastitis by CMT and Electroconductivity in the Camel*; Higher National Veterinary School: Algiers, Algeria, 2014. (In French)
7. Aljumaah, R.S.; Almutairi, F.F.; Ayadi, M.; Alshaikh, M.A.; Aljumaah, A.M.; Hussein, M.F. Factors influencing the prevalence of subclinical mastitis in lactating dromedary camels in Riyadh Region, Saudi Arabia. *Trop. Anim. Health Prod.* **2011**, *43*, 1505–1610. [CrossRef] [PubMed]
8. Abdurahman, O.A.S. The detection of subclinical mastitis in the Bactrian camel (Camelus bactrianus) by somatic cell count and California mastitis test. *Vet. Res. Commun.* **1996**, *20*, 9–14. [CrossRef] [PubMed]
9. Saleh, S.K.; Faye, B. Detection of subclinical mastitis in dromedary camels (Camelus dromedaries) using somatic cell counts, California mastitis test and udder pathogen. *Emir. J. Food Agric.* **2011**, *23*, 48–58.

biology and life sciences forum

MDPI

Proceeding Paper

Assessment of Contamination of Raw Camel Milk by *Listeria* spp. and *Staphylococcus* spp. †

Sara Lezzoum-Atek [1,2,*], Chahrazed Belhout [1], Hadir Bouchenafa [1] and Leila Bouayad [1]

[1] HASAQ Laboratory, High National Veterinary School, Issad Abbes Avenue, Oued Smar, El Harrach, Algiers 16270, Algeria; c.belhout@etud.ensv.dz (C.B.); h.bouchenafa@etud.ensv.dz (H.B.); l.bouayad@ensv.dz (L.B.)
[2] Faculty of Sciences, University of Ben Youcef Benkhedda, Algiers 16000, Algeria
* Correspondence: saralezzoum@gmail.com
† Presented at the 10th International Seminar of Veterinary Medicine: Camelids in Algeria & Maghreb, Constantine, Algeria, 20–21 December 2022.

Abstract: Camel milk is a valuable food choice, representing a primary need in the diet of people in dry zones. Camel milk is rich in nutrients, which makes it a favorable environment for the development of microorganisms. *Staphylococcus* and *Listeria* are significant opportunistic pathogens in humans, dairy cattle, and camels. The presence of these bacteria could present a potential public health issue. In the present study, 20 milk samples collected from camel farms in M'sila were investigated for the presence of *Staphylococcus* spp. and *Listeria* spp. *Staphylococcus* enumeration, and a search for *Listeria* spp. was performed according to the recommendations of ISO 6888-1(2004) and ISO 11290-1(2017) methods, respectively. The results show a contamination prevalence of 62% of *staphylococcus* spp. with an estimated average bacterial load of 2.7.102 cfu/mL, while for *Listeria* spp., only three samples were positive, with a prevalence of 14.28%. For each species, identification using API *Listeria* strips confirmed the presence of *Listeria grayi*, *Listeria innocua*, and *Listeria seeligeri* species, but no Listeria monocytogenes were recovered in these samples. According to the current results, we could conclude that the percentage of contamination with *Staphylococcus* in the tested camel milk samples was relatively high compared to the level of contamination with Listeria. There is no Algerian regulation setting microbiological criteria for raw camel milk; however, these results suggest that the hazard to the consumer cannot be excluded.

Keywords: camel's milk; *Staphylococcus* spp.; *Listeria* spp.; prevalence; foodborne pathogens

Citation: Lezzoum-Atek, S.; Belhout, C.; Bouchenafa, H.; Bouayad, L. Assessment of Contamination of Raw Camel Milk by *Listeria* spp. and *Staphylococcus* spp.. *Biol. Life Sci. Forum* **2023**, 22, 9. https://doi.org/10.3390/blsf2023022009

Academic Editors: Amira Leila Dib, Said Boukhechem, Hithem Bougherara and El-Hacene Bererhi

Published: 31 March 2023

1. Introduction

Camels are important to the lifestyle of several communities, particularly those of the Middle East and Africa [1]. Camel milk is one of the main components of diet in these arid and semiarid zones, where feed resources are scarce [2]. Camel's milk is rich with vitamins including B1, B2, and C [3]. Compared to cow's milk, camel milk contains three to five times more vitamin C, which makes it an important part of the diet in arid areas where accessibility to green foods is limited [4]. The camel milk is traditionally consumed predominantly in its raw state without any heat treatments [5], and in general the milk secreted by healthy cells has historically been considered sterile [6]. Its high content of antibacterial factors (Lactoferrin, Lactoperoxidase, and Lysozyme) gives it a particular capacity to be stored for a few days at relatively high temperatures (around 25 °C). However, it can be contaminated by pathogenic microorganisms of endogenous origin (*Staphylococci*, *Streptococci*, or *Escherichia coli*), following excretion from the udder of an infected animal or exogenous origin *(Bacillus*, *Clostridium*, *Micrococcus*, *Salmonella*, and *Listeria*) through direct contact with infected herds or through the environment (e.g., utensils, personnel) [7]. Microbiological criteria for raw cow's milk are well defined by regulations, yet there are no such criteria for camel milk, and published data on its bacterial hazards are also scarce.

Therefore, the objective of this research was to enumerate *Staphylococcus* spp. and isolate *Listeria* spp. from raw camel milk. These bacteria are often associated with the raw milk of other animal species, and have been directly linked to human and animal infections.

2. Material and Methods

Twenty camel milk samples were gathered from different farms in M'sila, which is a steppic zone located 200 km south of Algiers. The raw milk samples were obtained and stored in labeled screw-top bottles and were kept in an ice box under cold conditions during their direct transfer to the laboratory. *Staphylococci* were enumerated by using the spread plate technique in accordance with the EN ISO 6888–1 (2004). Presumptive staphylococcal colonies on Baird-Parker agar were confirmed using conventional methods, including colony morphology (black colony surrounded by a light halo) and catalase testing. The investigation of *Listeria* spp. was conducted according to the EN ISO 11290-1 method (ISO, 2017). Species identification of *Listeria monocytogenes* was performed based on the characteristic appearance of colonies on Aloa agar (bioMerieux) (green surrounded by a slight halo). All other species (not surrounded by a halo) were identified using API *Listeria* strips (BioMérieux).

3. Results and Discussion

This study assessed the prevalence of *Staphylococcus* spp. and *Listeria* spp. in raw camel milk. The overall prevalence of *Staphylococcus* spp. was 62% (Figure 1), with the average level of staphylococcal contamination in positive samples reaching 3×10^2 CFU/mL.

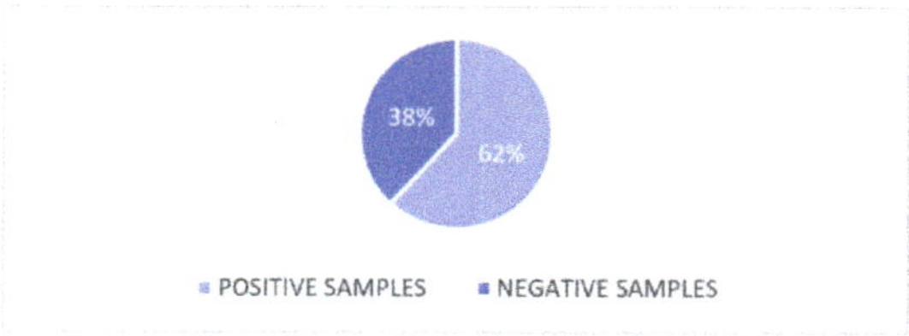

Figure 1. Prevalence and contamination level of *staphylococcus* spp. in raw camel milk.

The rates of contamination found were lower than those reported by Abera et al. [8] (89.8%), who investigated raw milk from 126 camels. In another survey on the microbiological quality of camel's milk in Algeria, Mosbah et al. [9] reported an overall prevalence of staphylococci of 2.8%, which is much lower than that identified in our results for the prevalence of *Listeria* spp. in raw camel milk (Table 1). *L. monocytogenes* was not detected, which corroborates the results of Debbouz et al. [10] and Ghardaïa and Ibrahim Rahimi et al. [11], who found all camel milk samples from camel breeding farms were negative for *Listeria monocytogenes*. The results of this study could be related to a lack of compliance with good production and hygiene practices during milking.

Table 1. Prevalence of *Listeria* species in raw camel milk.

Samples (N = 20)	*Listeria* Species			
	L.grayi N (%)	*L.innocua* N (%)	*L.seeligeri* N (%)	*L.monocytogenes* N (%)
Prevalence	1 (33.3%)	1 (33.3%)	1 (33.3%)	0

N: number, %: prevalence.

4. Conclusions

The Algerian regulation does not consider camel milk in the microbiological criteria set for foods. Nevertheless, the results show high contamination by *Staphylococcus* spp. and significantly lower contamination by *Listeria* spp., which suggests that the danger for the consumer cannot be excluded. The information obtained from this study could be useful for epidemiological studies on *Staphylococcus* spp. and *Listeria* spp. for public health considerations.

Author Contributions: Conceptualization, S.L.-A., C.B., H.B. and L.B.; methodology, S.L.-A.; validation, S.L.-A., C.B. and L.B.; investigation, S.L.-A., C.B., H.B. and L.B.; writing—original draft preparation, S.L.-A., C.B. and L.B.; writing—review and editing, S.L.-A., C.B. and L.B.; supervision, L.B. All authors have read and agreed to the published version of the manuscript.

Funding: This research received no external funding.

Institutional Review Board Statement: Not applicable.

Informed Consent Statement: Not applicable.

Data Availability Statement: Not applicable.

Conflicts of Interest: The authors declare no conflict of interest.

References

1. Mirkena, T.; Walelign, E.; Tewolde, N.; Gari, G.; Abebe, G.; Newman, S. Camel production systems in Ethiopia: A review of literature with notes on MERS-CoV risk factors. *Pastoralism* **2018**, *8*, 1–17. [CrossRef] [PubMed]
2. Farah, Z. *Camel Milk: Properties and Products*; Swiss Centre for Development Cooperation in Technology and Management: St. Gallen, Switzerland, 1996.
3. Singh, R.; Mal, G.; Kumar, D. Camel Milk: An Important Natural Adjuvant. *Agric. Res.* **2017**, *6*, 426–434. [CrossRef]
4. Ereifej, K.I.; Alu'datt, M.H.; Alkhalidy, H.A.; Alli, I.; Rababah, T. Comparison and characterisation of fat and protein composition for camel milk from eight Jordanian locations. *Food. Chem.* **2011**, *127*, 282–289. [CrossRef]
5. Handling, Preservation and Utilization of Camel Milk and Camel Milk Products in Shinile and Jijiga Zones, Eastern Ethiopia. Available online: http://www.lrrd.org/lrrd19/6/seif19086.htm (accessed on 30 May 2022).
6. Swelum, A.A.; El-Saadony, M.T.; Abdo, M.; Ombarak, R.A.; Hussein, E.O.S.; Suliman, G. Nutritional, antimicrobial and medicinal properties of Camel's milk: A review. *Saudi. J. Biol. Sci.* **2021**, *28*, 3126. [CrossRef] [PubMed]
7. Brisabois, A.; Lafarge, V.; Brouillaud, A.; de Buyser, M.L.; Collette, C.; Garin-Bastuji, B.; Thorel, M.F. Pathogenic organisms in milk and milk products: The situation in France and in Europe. *Rev. Sci. Tech.* **1997**, *16*, 452–471. [CrossRef] [PubMed]
8. Abera, T.; Legesse, Y.; Mummed, B.; Urga, B. Bacteriological quality of raw camel milk along the market value chain in Fafen zone, Ethiopian Somali regional state. *BMC Res. Notes* **2016**, *9*, 1–6. [CrossRef] [PubMed]
9. Qualité Microbiologique du lait de Chamelle (*Camelus dromedarius*) Élevée en Systeme Semi Intensif dans la Localite de Ghardaïa. Revue des Bio-Ressources. Available online: https://journals.univ-ouargla.dz/index.php/RBR/article/view/1371 (accessed on 1 June 2022).
10. Debouz, A.; Guerguer, L.; Hamid Oudjana, A.; Hadj Seyd, A.E.K. Etude comparative de la qualité physico-chimique et microbiologique du lait de vache et du lait camelin dans la wilaya de Ghardaïa. *Rev. ElWahat Pour Les Rech. Et Les Etudes.* **2014**, *7*, 10–17.
11. Rahimi, E.; Momtaz, H.; Behzadnia, A.; Baghbadorani, Z.T. Incidence of *Listeria* species in bovine, ovine, caprine, camel and water buffalo milk using cultural method and the PCR assay. *Asian Pac. J. Trop. Dis.* **2014**, *4*, 50–54. [CrossRef]

biology and life sciences forum

Proceeding Paper

Teeth Follow-Up throughout the Life of the Dromedary: Zootechnical and Veterinary Importance †

Mabrouk Mouldi Seddik

Livestock and Wildlife Laboratory, Institute of Arid Regions, Medenine 4119, Tunisia; mabrouk.seddik16@gmail.com

† Presented at the 10th International Seminar of Veterinary Medicine: Camelids in Algeria & Maghreb, Constantine, Algeria, 20–21 December 2022.

Abstract: The dromedary camel is a domestic animal in arid and desert regions. Its oral cavity has evolved to allow the ingestion of desert plants, often not consumed by other species. Indeed, the teeth of this species present anatomical peculiarities specific to pseudo-ruminant animals by their forms, their arrangements, their formulas, and their kinetics of eruption and wear. With 22 deciduous and 34 permanent teeth, including 2 incisors and 2 canines in the upper jaw and 6 incisors and 2 canines in the lower jaw, estimating the age of a dromedary remains difficult for most clinicians. This accurate age determination is very important not only for breeders and young promoters when purchasing animals but also for the clinician when performing veterinary or zootechnical examinations. This work was carried out in a herd of 70 camels belonging to the Arid Lands Institute (IRA Tunisia) and is based on a monthly and annual follow-up and examination of the morphology of the teeth (eruptions, replacements, and wear) throughout the life of the dromedary.

Keywords: teeth; estimation age; dromedary camel

check for updates

Citation: Seddik, M.M. Teeth Follow-Up throughout the Life of the Dromedary: Zootechnical and Veterinary Importance. *Biol. Life Sci. Forum* **2023**, 22, 10. https://doi.org/10.3390/blsf2023022010

Academic Editors: Amira Leila Dib, Said Boukhechem, Hithem Bougherara and El-Hacene Bererhi

Published: 4 April 2023

1. Introduction

In camel breeding, the follow-up of the animals is not so easy, and the determination of the age is very important in veterinary semiology. The dentition remains of great importance in the estimation of the camel age by the breeders, so like all other mammals, camels have temporary or permanent teeth. This species has 22 deciduous teeth and 34 permanent teeth [1,2].

2. Material and Methods

This work was based on a monthly and annual follow-up and examination of the morphology of the teeth (eruptions, replacement, and wear) throughout the life of the dromedary. It was carried out in the IRA herd of 70 dromedaries in a semi-extensive system grazing halophilic plants during the day (6 to 7 hours) and receiving 2 kg of mixed feed (40% crushed barley, 40% olive cake, 15% bran, and 5% CMV) in the afternoon. The observed camels were in good health, treated monthly against ectoparasites, dewormed annually in early spring, and raised in semi-extensive systems owned by the Arid Lands Institute (Medenine, Tunisia). This survey involved different age groups, including young animals from birth to 4 years, adults from 5 to 20 years, and cull animals from 21 to 30 years, in order to facilitate the estimation of the age of this species.

3. Results and Discussion

The dental formula of the young dromedary is composed of a total number of 22 temporary or deciduous teeth divided into incisors: 2/6; canines 2/2; premolars 6/4. In contrast, the number of adult or replacement teeth is 34 [3], and their structure is as follows: In the upper jaw, there are 2 incisors, 2 canines, 6 premolars, and 6 molars; While in the lower jaw, there are 6 incisors, 2 canines, 4 premolars and 6 molars. The size of the deciduous teeth was smaller than that of the permanent teeth. A clear difference was observed when they were present with the permanent teeth on the animal's jaw [2]. The determination of the age of the camel was based on the examination of the incisors and canines of the upper and lower jaws. The dromedary camel is born without teeth. After one week of age, the incisor pincers appear, and after two weeks, the pair of temporary intermediate incisors appear. In addition, after one month of age, the corner incisors appear. Between 2 and 4 months of age, the canines of the lower jaw appear, while those of the upper jaw appear between 8 and 12 months of age. In fact, at the end of the first year of age, the species completes its milk teeth. At 3.5 years of age, there is a very important wear to the deciduous incisors; all incisors have a pronounced neck. Thus, the camel begins to replace its deciduous teeth with permanent teeth when it reaches the age of 4 years. The permanent teeth are larger, longer, and darker in color and do not have the well-defined neck that joins the temporary teeth. Therefore, by four to five years of age, both pincers fall out, and two new incisor pincers emerge. At five to six years, the second incisors (intermediate teeth) fall out, and two new incisors emerge. From six to seven years of age, the third incisors (corners) fall out, and two new incisors appear. Thus, at the age of seven, the camel completes the appearance of permanent incisors. At the age of eight, the permanent canines appear, and the shape of the incisors is completed; this is called "full mouth". In the 10-year-old camel, all permanent canine teeth (upper jaw corners, all first premolars of the permanent teeth) are well developed. In addition, a little wear is observed on the incisor pincers. At the age of 12 years, the incisors are more worn with a bi-angular pincer-like surface. At age 15, the surface of the pincer teeth is semicircular, and the intermediate incisors are bi-angular. At the age of 17, a black dot can be seen on the surface of the claw teeth, called a radical star. At the age of 18, the surface of the claws appears oval, and the intermediate incisors are semicircular. At the age of 20, wear of the canines on the lower jaw is observed with a bi-angular surface shape. At age 22, wear of all teeth is observed with a spacing of the incisor teeth; in addition, the surface of the claws and intermediate incisors appears circular; this observation has not been reported in other studies [4–6]. In addition, at the age of 24–28 years, the incisors appear triangular with the persistence of the black pulp cavity.

4. Conclusions

The monitoring and examination of the morphology of camel teeth throughout its life require good restraint of the animal, which is not easy for this species. However, this age estimate remains very important for any breeding management where the productive life of the animal rarely exceeds 25 years. The wearing kinetic depends mainly on the used breeding system and the pastoral plant cover.

Funding: This research received no external funding. However, my participation in the Xth SIMV has been partially supported by the WELLCOME TRUST "COBRA-NGaV" Project: Developing the next generation of antivenoms to improve the management of cobra poisonings in LMI countries. The contents of this document are the sole responsibility of «Institut des Régions Arides de Médenine, Tunisia» partner.

Institutional Review Board Statement: Not applicable.

Informed Consent Statement: Not applicable.

Data Availability Statement: Data is available upon request.

Acknowledgments: My special thanks are extended to the staff of the Livestock and Wildlife Laboratory, Institute of Arid Regions of Medenine, especially Boughuedima Mohsen and Kchira Belguecem, for their help during camel teeth examination.

Conflicts of Interest: The author declares no conflict of interest.

References

1. Food and Agriculture Organization of the United Nations (FAO). Camels; llamas and alpacas. In *A Manual for the Primary Animal Health Care Workers*; FAO: Rome, Italy, 1994; Chapter 7.
2. Faye, B. *Guide de l'élevage du dromadaire*; Sanofi: Paris, France, 1997.
3. Camel Teeth: Everything You Need to Know. Available online: https://a-z-animals.com/blog/camel-teeth-everything-you-need-to-know/ (accessed on 29 December 2021).
4. Ghadry, G. *Horses and Camels*; Faculty of Agriculture, Aleppo University: Deir Ezzor, Syria, 1983.
5. Misk, N.A.Y.; Youssef, H.A.; Semieka, M.M.; El-Khabery, A.H. History of definition in camels. In Proceedings of the XXXVII International Congress of the World Association for the History of Veterinary Medicine & XII Spanish National Congress on the Veterinary History, León, Spain, 21–24 September 2006; pp. 535–540.
6. Chaurasia, S.; Tiwari, Y.; Pandey, A.; Kumar, V.; Malik, M.R. Biometry of mandible of camel (Camelus dromedarius). *Int. J. Mol. Biol.* **2010**, *1*, 25–28.

biology and life sciences forum

MDPI

Proceeding Paper

Main Reasons for the Seizure of Meat and Offal in Slaughterhouses in the Region of Adrar (Algeria) †

Meriem Bouzenzana [1,*] and Naima Hadjab [2]

1 Departement of Animals Productions, Institute of Veterinary Sciences, El Khroub, University Constantine 1, Constantine 2500, Algeria
2 Department of Veterinary and Agriculture Sciences, Hadj Lakhdar University of Batna 1, Batna 05000, Algeria; hadjab.naima@univ-batna.dz
* Correspondence: bouzenzana.meriem@umc.edu.dz
† Presented at the 10th International Seminar of Veterinary Medicine: Camelids in Algeria & Maghreb, Constantine, Algeria, 20–21 December 2022.

Abstract: The main of this study is to determine the nature and frequency of the different reasons for the seizure of mean and offal in the camel compared to the other ruminants slaughtered under the same conditions. This study is a synthesis of the slaughterhouse records that are part of the activity of the veterinary inspection office of the agricultural services of the wilaya of Adrar during 2017. Camels accounted for 10.4% of the total meat produced at the slaughterhouse in 2017, after the sheep, with 77.8%. The percentage of seizures for meat in camel is 3.12%. The highest weight of seized meat is observed in sheep, with 3665 kg, then camel, with 150 kg. Traumatized meat is the predominant reason for seizures, with 3.12% (150 kg) and 81.81% (115 kg) for camel and sheep, respectively. The dromedary is in the second class for the seizure of red offal after sheep, with an estimated loss of 316.3 kg, of which 88.88% of seizures are represented by lung lesions. Pneumonia is the greatest reason for offal seizure at 1.1%, followed by pulmonary hydatidosis at 0.36%. The dominant cases lead to considerable losses of animal protein. The loss of protein in a country that lacks sufficient sources shows how worrying seizures are.

Keywords: camel; slaughterhouse; seizure; meat; offal

Citation: Bouzenzana, M.; Hadjab, N. Main Reasons for the Seizure of Meat and Offal in Slaughterhouses in the Region of Adrar (Algeria). *Biol. Life Sci. Forum* **2023**, *22*, 12. https://doi.org/10.3390/blsf2023022012

Academic Editors: Amira Leila Dib, Said Boukhechem, Hithem Bougherara and El-Hacene Bererhi

Published: 4 April 2023

1. Introduction

In Algeria, Camelin breeding plays an economic, social, and cultural role and is part of the country's animal production development strategy [1]. It is present in 17 wilayas, 8 Saharans, and 9 steppics [2].

At the national level, Adrar wilaya ranked second with an average size that stabilized around 41 thousand head between 2000 and 2015 year, after Tamanrasset [1]. Indeed, camel is considered an important source of animal protein for Saharan populations [3,4], cited by [5]. The main of this study is to assess the reasons for the seizure of meat and offal of camels during the 2017 year in slaughterhouses in comparison with the other ruminants and to assess the associated economic loss for the breeder.

2. Material and Methods

The wilaya of Adrar occupies a large part of the Algerian southwest; it covers an overall area of 17.97% of the national territory [6]. The climate is desert-type. This wilaya has 8 slaughterhouses. Four animal species were the subject of our study, namely cattle, sheep, goats, and camelins, which are slaughtered in each slaughterhouse belonging to the wilaya of the study.

The statistical data presented by this study were obtained by a synthesis of the monthly stock records of slaughterhouses in the wilaya during 2017.

The total number of animals inspected shall be cattle (1037 heads, 18,482 tons of meat), sheep (77,753 heads and 1,509,901 tons of meat), camels (10,425 heads and 198,766 tons of meat), and goats (10,710 heads and 1875 tons of meat).

The percentage of sanitary seizure was calculated for each pathological reason, both for carcasses and for offal.

3. Results and Discussion

3.1. Proportion of Camelin Meat Slaughtered in Relation to Other Red Meat

The number of dromedaries slaughtered varies according to market demand. Camels accounted for 10.4% of the total meat produced at the slaughterhouse in 2017. It is in second place after ovine meat, which represents the highest proportion, with an average of 77.8%. This is followed by the goat, which represents a rate of 11%, and the last is beef, with a timid percentage of 1%. This level remains high, following the strong demand by consumers for this meat for dietary and consumption reasons [5]. The number of camels slaughtered is around 10,424, or a monthly average of 867.

3.2. Percentage of Meat Seized by Reason

The overall percentage of meat seizure observed during our study is high for sheep, with 90.62% (29 carcasses out of 32), followed by beef, camel, and goat, with 3.12% for each one. The total weight of seized meat is 527 kg. The highest weight of seized meat is observed in sheep with 3665 kg, then by camel with 150 kg, and finally with cattle with 10 kg.

This study showed that the majority of meat seizures are the result of various accidents, with 34.35% (275 kg) (catching, transport, and especially traffic accident victims, especially for camels), expressed by the predominance of seizures of traumatized meat with, respectively, 3.12% (10 kg), 3.12%(150 kg), and 81.81% (115 kg) for cattle, camel, and sheep.

3.3. Percentage of Offal Seized by Reason

For offal, the highest overall percentage of seizure is for red offal of sheep, with a percentage of 97.34%, followed by red offal of camels at 1.15%. Compared with camel, liver hydatidosis in sheep represents the highest rate of red offal seizure (61.12% vs. 0.33%), followed by lung hydatidosis (47.7% vs. 0.36%), liver abscess (34.69% vs. 0.11%), and lung abscess (34.29% vs. 0.13 %). While in camel, pneumonia is the most reason for offal seizure (1.1% vs. 15.02%), 88.88% of seizures are represented by lung lesions. In the other species, the lung lesions are, respectively, 79.64 %, 70.64%, and 53.84% in goats, sheep, and cattle.

The economic losses associated with the seizure of red offal were remarkable, with sheep occupying the first position with 4404.2 kg. The dromedary is in the second class with an estimated loss of 316.3 kg. It is noted that the total weight of the lung seized represents two-thirds of the weight of the liver seized in sheep (30,059 kg vs. 13,983 kg), and ten times greater in dromedary (2873 kg vs. 29 kg).

4. Conclusions

The purpose of this study is to determine the nature and frequency of the different seizure reasons for meat and offal in the camel compared to the other ruminants slaughtered under the same conditions.

A comparison between the animals studied has the objective of drawing the conclusion of the profitability of the camel compared to other species; in addition, it is known for its resistance to thirst, heat, and protein undernutrition in a very harsh environment; it is also very resistant to a large number of pathologies.

The dominant motives lead to considerable losses of animal protein.

Author Contributions: Conceptualization, M.B.; methodology, M.B.; software, M.B.; validation, M.B.; formal analysis, M.B.; investigation, M.B. and N.H.; resources, M.B. and N.H.; data curation, M.B.; writing—original draft preparation, M.B.; writing—review and editing, M.B.; visualization, M.B. and N.H.; supervision, M.B.; project administration, M.B.; funding acquisition, M.B. All authors have read and agreed to the published version of the manuscript.

Funding: This research received no external funding.

Institutional Review Board Statement: Not applicable for studies not involving humans or animals.

Informed Consent Statement: Informed consent was obtained from all subjects involved in the study.

Data Availability Statement: We declare that no new data were created; data are unavailable until now. We will create it soon.

Acknowledgments: The authors would like to thank the director of agricultural services of the wilaya of Adrar, which gave us access to the data of the monthly reports of slaughterhouses of the wilaya, which are part of the activity of the veterinary inspection.

Conflicts of Interest: The authors declare no conflict of interest.

References

1. Meguellati-Kanoun, A.; Saadaoui, M.; Kalli, S.; Kanoun, M.; Huguenin, J.; Benidir, M.; Benmebarek, A. Localisation et distribution spatiotemporelle des effectifs de dromadaires en Algérie. *Livest. Res. Rural. Dev.* **2018**, *30*, 1–8.
2. Aissa, B. Le dromadaire en Algérie. *Options Méditerranéennes* **1989**, *2*, 19–28.
3. Adamou, A. L'élevage camelin en Algérie: Quel type pour quel avenir? *Sci. Et Changements Planétaires/Sécheresse* **2008**, *19*, 253–260.
4. Faye, B. L'élevage des grands camélidés: Vers un changement de paradigme. *Rencontres Autour Des Rech. Sur Les Rumin.* **2019**, *16*, 345–348.
5. Sadoud, F.M.; Nefnouf, F.; Hafaoui, F. La viande cameline dans deux régions du Sud Algérien. Viandes Prod. *Carnés Référence* **2019**, *3532*, 1–11.
6. Bouhous, A.; Aissi, M.; Harhoura, K.H. Etude des Ixodidae chez le dromadaire dans le sud algérien, région d'Adrar. *Ann. Méd. Vét.* **2008**, *152*, 52–58.

biology and life sciences forum

MDPI

Proceeding Paper

Cytological and Immunohistochemical Study of the Dromedary lymph Nodes in Algeria †

Mohamed Amine Fares [1,*], Tarek Khenenou [1] and Djallal Eddine Rahmoun [2]

1 Laboratory of Sciences and Techniques of Living, Institute of Agronomic and Veterinary Science, Taoura 41000, Algeria; tarek.khenenou@univ-soukahras.dz
2 Laboratory of Animal Production, Biotechnologies and Health (PABIOS), Institute of Agronomic and Veterinary Science, Taoura 41000, Algeria; d.rahmoun@univ-soukahras.dz
* Correspondence: m.fares@univ-soukahras.dz

† Presented at the 10th International Seminar of Veterinary Medicine: Camelids in Algeria & Maghreb, Constantine, Algeria, 20–21 December 2022.

Abstract: The aim of this research is to make an investigation about the cellular population of the dromedary lymph nodes in the region of El Oued in Algeria in order to identify the cytological structure of these organs; a classic histological staining technique had to be performed in order to identify the cellular population in each compartment of the organ. Moreover, it was necessary to make an immunohistochemical staining technique using monoclonal ant bodies in order to identify the localization of both T & B lymphocytes. The obtained results revealed that the location of both T and B lymphocytes in the dromedary's lymph node is identical to the general organization of lymph nodes of other mammalian species.

Keywords: azure II eosin; CD22; CD3; dromedary; immunohistochemistry; lymph node; lymphocyte

check for updates

Citation: Fares, M.A.; Khenenou, T.; Rahmoun, D.E. Cytological and Immunohistochemical Study of the Dromedary lymph Nodes in Algeria. *Biol. Life Sci. Forum* **2023**, 22, 8. https://doi.org/10.3390/blsf2023022008

Academic Editors: Amira Leila Dib, Said Boukhechem, Hithem Bougherara and El-Hacene Bererhi

Published: 31 March 2023

1. Introduction

The dromedary's immune system is one of the most mysterious topics for researchers in the fields of morphology and immunology; moreover, the lack of information about the dromedary immune system led us to conduct the current research, which is based on one of the most important lymphoid organs; the lymph nodes of one-humped camel (camelus dromedarius) from the region of El Oued in southeastern Algeria, in which we aimed to make a full cytological and immunohistochemical study of the axillary, and the mesenteric lymph nodes, in order to highlight the structure and the organization of functional compartments of these organs.

2. Material and Methods

The lymph nodes have undergone meticulous preparation, which starts with the removal of the adipose tissue covering these organs. This process required the use of a scalpel. The sampled organs were then immersed in a 10% formalin solution for 24–48 h for fixation; additional fixation was carried out at room temperature in a 10% formalin solution for 10 to 14 days for the fragments that were meant to be stained using Azur II eosin technique, whereas the identification and the localization of T and B lymphocytes were carried out using immunohistochemistry technique with both anti-CD3 and anti-CD22 monoclonal antibodies to elucidate T and B lymphocytes location.

3. Results and Discussion

The results of the cytological study of the lymph nodes using Azur II eosin stains revealed that the lymph nodes parenchyma is composed of several cellular populations, in which the lymphocytes were the major population, followed by the macrophages, plasma cells, reticulocytes, and granulocytes such as eosinophils, basophils, and neutrophils (Figure 1). Moreover, it was possible to elucidate that the ratio of lymphocytes in the active follicles and in the paracortical zone was higher than the other functional zones. Furthermore, they presented a diameter of 8–12 μm, with a very dense, rounded nucleus, while the large lymphocytes presented a diameter of 12–15 μm. Whereas macrophages were recognizable by a voluminous, clear, and well-defined cytoplasm. Plasma cells were found to be spreading along the medullary zone, and the dromedary parenchyma was distinguished by a higher rate of reticular cells, plasma cells, and macrophages (Figure 2). The study of the morphology of lymph nodes of the dromedary in the region of El Oued allowed us to highlight a distinct histological structure formed by several conglomerates consisting of a different number of small nodules scattered within the adipose tissue. Our results indicate that the parenchyma of the lymph nodes in a dromedary has a (compartmentalized) lobular structure that corresponds to the results of the researcher [1]. The immunohistochemical results showed that the B lymphocytes were located in the center of the germinal center of the follicles of the lymph nodes, whereas the T lymphocytes were located in the paracortical zone (Figures 2 and 3). Nearby identical results were found in the spleen [2].

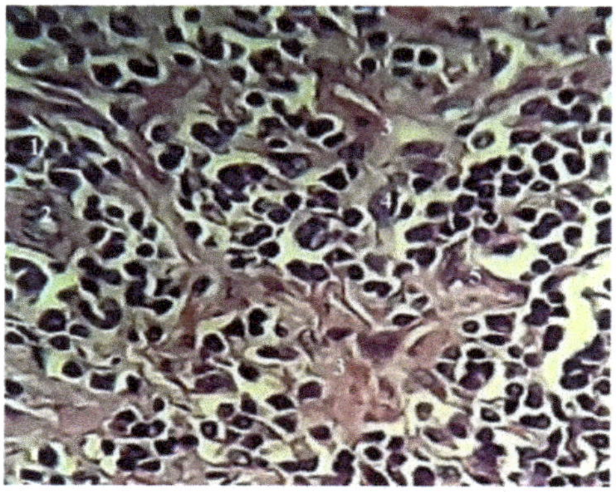

Figure 1. Histological section of the mesenteric lymph node of the dromedary, Azure II eosin staining, X100, 1—Large lymphocyte, 2—Blood vessel, 3—Reticulocyte, 4—Granulocyte, 5—Plasmocyte.

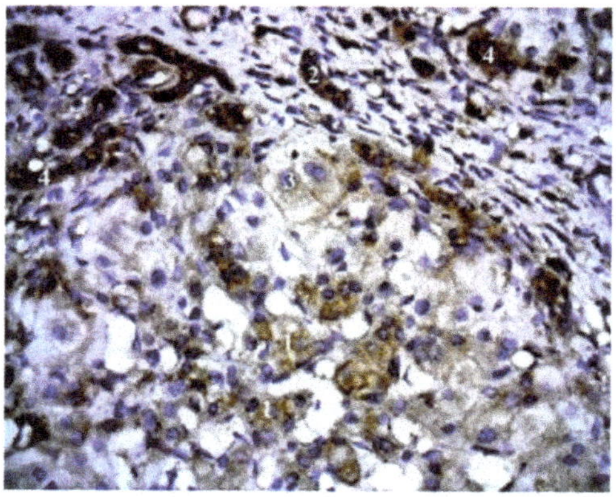

Figure 2. Anti-CD22 immunohistochemical reaction of the mesenteric lymph node of dromedary X40, Positive reaction. 1—T lymphocyte, 2—Cortical zone, 3—Cortico-medullary zone, 4—Exaggerated reaction to the T lymphocyte.

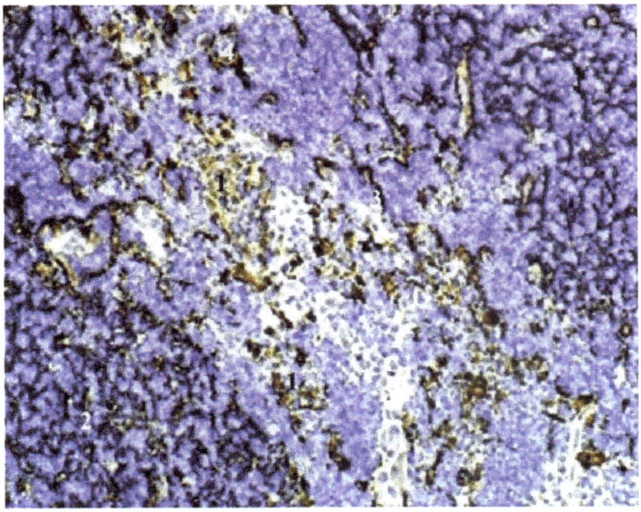

Figure 3. Anti-CD3 immunohistochemical reaction of the mesenteric lymph node of dromedary X40, Positive reaction. 1—B lymphocyte, 2—Medullary zone, 3—Cortico-medullary zone, 4—Exaggerated reaction to the B lymphocyte.

4. Conclusions

Cytological and immunohistochemical techniques were used to supplement the microscopic examination of the lymph node of the one-humped camel (camelus dromedarius), using azure II eosine staining technique and immunochemistry using monoclonal antibodies; CD3 and CD22, in order to highlight the location of T and B lymphocytes, and from the obtained results, it is now possible to say that all mammals share the same lymphocyte location, in which the B lymphocytes are located in the germinal center of the primary follicles, while the T lymphocytes are located in the paracortical zone, whereas the B lymphocytes were located in the follicles of both the cortical and the medullary zone and the T lymphocytes were located exclusively in the para-cortical region.

Author Contributions: Conceptualization, T.K., D.E.R. and M.A.F.; methodology, M.A.F.; software, M.A.F.; validation, M.A.F., T.K. and D.E.R.; formal analysis, M.A.F.; investigation, M.A.F.; resources, M.A.F.; data curation, M.A.F.; writing—original draft preparation, M.A.F.; writing—review and editing, M.A.F.; visualization, M.A.F.; supervision, T.K. and D.E.R.; project administration, T.K.; funding acquisition, M.A.F. All authors have read and agreed to the published version of the manuscript.

Funding: This research received no external funding.

Institutional Review Board Statement: The animal study protocol was approved by the Institutional Review Board of the institute of agriculture and veterinary sciences in Taoura, University of Souk Ahras, Algeria.

Informed Consent Statement: Not applicable.

Data Availability Statement: No new data were created or analyzed in this study. Data sharing is not applicable to this article.

Acknowledgments: The authors would thank Yassine Ladjailia (Laboratory Associate in the Institute of Agriculture and Veterinary Sciences, University of Souk Ahras) for the comments that greatly improved the manuscript.

Conflicts of Interest: The authors declare no conflict of interest.

References

1. Rahmoun, D.; Lieshchova, M.; Fares, M. Morphological and radiological study of lymph nodes in dromedaries in Algeria. *Regul. Mech. Biosyst.* **2020**, *11*, 330–337. [CrossRef]
2. Amine, F.M.; Tarek, K.; Eddine, R.D. Anatomo-topographic and histo-cytological study of dromedary's spleen in Algeria. *Folia Morphol.* **2023**, *82*, 137–146.

biology and life sciences forum

MDPI

Proceeding Paper

Diaphragm Bone in Dromedary (*Camelus dromedarius* L., 1758): Anatomy and Investigation Using Computed Tomography Imaging †

Faiza Tekkouk-Zemmouchi [1], Baaissa Babelhadj [2,3], Rania Ridouh [1,*], Eric Betti [4], Abdelkader Adamou [3], Louiza Benhamza-Manssar [1], Archibald Poisson [5], Atika Benaissa [3], Kassem Altoama [4], Christian Tavernier [5] and Claude Guintard [4]

1 Laboratory of Health Management and Animal Production, Institute of Veterinary Sciences, El khroub, University Frères Mentouri of Constantine 1, Constantine 25000, Algeria
2 Department of Natural Sciences, Superior Normal School of Ouargla, Ouargla 30000, Algeria
3 Laboratory of Ecosystems Protection in Arid and Semi-Arid Zones, University of Kasdi Merbah Ouargla, Ouargla 30000, Algeria; adamoudz@yahoo.fr (A.A.)
4 Comparative Anatomy Unit, National Veterinary School of Food and Agrifood, Nantes Atlantique-Oniris, Gachet Street, Cs 40706, CEDEX 3, 44307 Nantes, France
5 S.a.r.l. IMAGE ET, Image ET, 101B Avenue André Bonnin, 35135 Chantepie, France
* Correspondence: anatomiste27@outlook.fr
† Presented at the 10th International Seminar of Veterinary Medicine: Camelids in Algeria & Maghreb, Constantine, Algeria, 20–21 December 2022.

Abstract: The aim of this work is to provide some anatomical data using a non-invasive method. Nine entire diaphragms of adult dromedaries were collected randomly at the slaughterhouse, and the bones prepared at the anatomy laboratory of Ouargla (Algeria). The CT examinations were performed on a 16-section CT device made by Siemens (Sensation 16, dedicated to the environment and the veterinary industry by Image-ET (Mordelles, France)). This bone is flat on one side and protruding on the other. The central part of the bone has an average HU value of −176 (−684 to 88), which clearly corresponds to the cancellous bone, and the external surface of the bone has an average HU value of 2320 (1979 to 2664), which corresponds to the compact bone. This study allows us to have a better understanding of the variability and structure of the dromedary diaphragm bone.

Keywords: diaphragm; bone; dromedary; anatomy; scan

check for updates

Citation: Tekkouk-Zemmouchi, F.; Babelhadj, B.; Ridouh, R.; Betti, E.; Adamou, A.; Benhamza-Manssar, L.; Poisson, A.; Benaissa, A.; Altoama, K.; Tavernier, C.; et al. Diaphragm Bone in Dromedary (*Camelus dromedarius* L., 1758): Anatomy and Investigation Using Computed Tomography Imaging. *Biol. Life Sci. Forum* **2023**, 22, 17. https://doi.org/10.3390/blsf2023022017

Academic Editors: Amira Leila Dib, Said Boukhechem, Hithem Bougherara and El-Hacene Bererhi

Published: 10 April 2023

1. Introduction

In arid regions, the dromedary is a domestic animal raised in the same way as other farm animals (cattle, sheep, goats, horses, etc.) for its production of milk and meat, and for its ability to bat [1]. Its hardiness in a low productivity environment, its milk, its meat, and its work are very appreciated by breeders, whose life depends on it in the desert [2], even if this species has long remained marginal for scientific studies [3]. The bone of the diaphragm in the dromedary has always been present (Figure 1). Few studies about this bone exist [4,5], and nearly no morphometric study has evaluated its characteristics. Namshir [5] just indicates that the diaphragm bone is 0.6–1.4 cm thick, 1.7–2.2 cm wide, and 3.3 cm long. The aim of this work is to provide some anatomical data using a non-invasive method, i.e., CT scanning. The scanner is widely used in human and veterinary medicine as a complementary examination. It allows 3D reconstructions of the organs. Here, it allows access to the internal structure of the bones without damaging them.

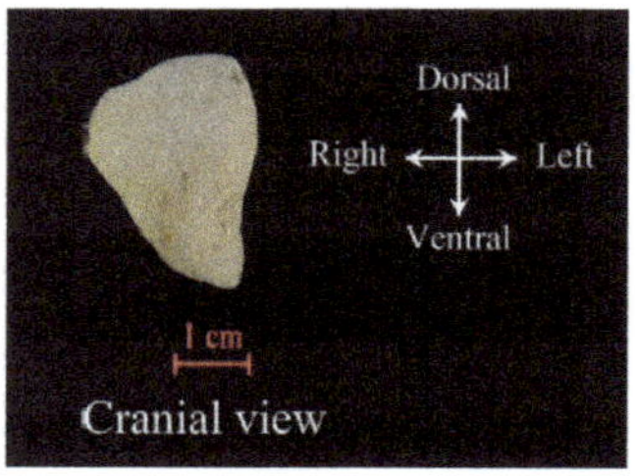

Figure 1. Dromedary diaphragm bone.

2. Materials and Methods

A total of nine entire diaphragms of the dromedary were collected randomly at a slaughterhouse and prepared at the anatomy laboratory of Ouargla. The diaphragm bones (Figure 2) were sent to France for CT scan analysis. CT examinations were performed on a 16-slice computed tomography device manufactured by Siemens (Sensation 16, dedicated to the veterinary environment and industry by Image-ET (Mordelles, France)). The CT scanning was performed with a thickness of 750 μm per acquisition. Herein, 3D reconstructions were made from 100 to 200 images according to the bone size.

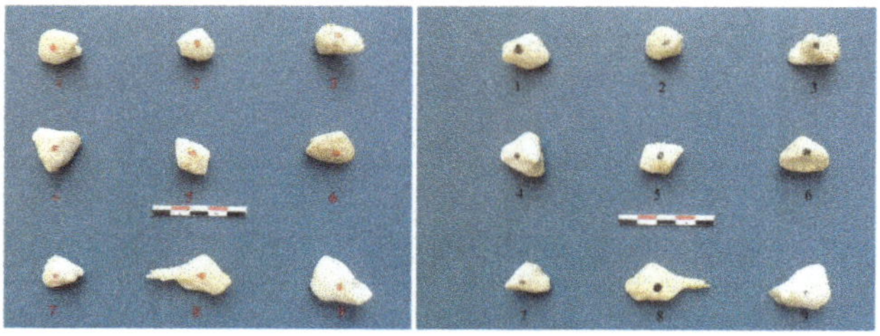

Figure 2. Cranial (**left side**) and caudal (**right side**) views of the nine diaphragm bones.

3. Results and Discussion

This bone is flat on one side and protuberant on the other. The central part of the bone has a mean HU value of −176 (−684 to 88), which corresponds clearly to the spongious bones, and the external surface of the bone has a mean HU value of 2320 (1979 to 2664), corresponding to a compact bone (Table 1). Few studies about this bone exist [4,5], and nearly no morphometric study presents its characteristics. These results are the first obtained on this bone using CT scans. No comparison studies have been published to our knowledge.

Table 1. Size (CC [thickness], PD [length] and LR [width]) in cm and CT scan density in Hounsfield Unit (UH).

Bone	CC (cm)	PD (cm)	LR (cm)	UH max (ext.)	UH max (int.)	UH min (int.)	UH mean (int.)
1	0.99	2.45	1.75	2664	584	−981	−60
2	1.00	1.55	1.78	2334	654	−1024	−65
3	1.09	1.61	2.39	1979	381	−1024	−263
4	1.40	2.38	2.45	2470	509	−1024	−684
5	0.89	1.78	2.43	2597	915	−1012	−58
6	1.24	1.92	2.65	2315	605	−1024	−274
7	1.06	1.95	2.21	2328	526	−1024	−325
8	0.81	2.22	4.49	2063	761	−1007	59
9	1.08	2.34	3.21	2127	938	−884	88

Table 1. *Cont.*

Bone	CC (cm)	PD (cm)	LR (cm)	UH max (ext.)	UH max (int.)	UH min (int.)	UH mean (int.)
Mean	1.06	2.02	2.59	2320	653	−1000	−176
Max	1.40	2.45	4.49	2664	938	−884	88
Min	0.81	1.55	1.75	1979	381	−1024	−684
SD	0.18	0.34	0.84	234	187	46	240

It consists of a central part with the marrow and a peripheral compact cortex. The size of the bone is 0.81–1.40 cm thick, 1.55–2.45 cm wide, and 1.75–4.49 cm long.

4. Conclusions

These preliminary results give us a better understanding of the variability and the structure of the dromedary diaphragm bone.

Author Contributions: Conceptualization, C.G. and B.B.; methodology, C.G.; software, K.A.; validation, C.T., A.P. and C.G.; formal analysis, E.B.; investigation, B.B.; resources, A.A., A.B. and L.B.-M.; data curation, R.R.; writing—original draft preparation, C.G.; writing—review and editing, F.T.-Z.; visualization, F.T.-Z.; supervision, C.G.; project administration, C.G.; funding acquisition, R.R. and F.T.-Z. All authors have read and agreed to the published version of the manuscript.

Funding: This research received no external funding.

Institutional Review Board Statement: Not applicable.

Informed Consent Statement: Not applicable.

Data Availability Statement: The works consulted are detailed in the bibliography.

Conflicts of Interest: The authors declare no conflict of interest.

References

1. Guintard, C.; Babelhadj, B. Morphotypes et force animale développée. Comparaison de deux populations de dromadaires algériens: La Sahraoui et la Targui (*Camelus dromedarius*, L.). In *Animal Source D'énergie: Enquêtes dans l'Europe Pré-Industrielle*; Guizard, F., Beck, C., Eds.; Presses Universitaires de Valenciennes: Valenciennes, France, 2018; pp. 133–147.
2. Faye, B. *Statut Nutritionnel du Bétail Dans la République de Djibouti*; Ministère de la Coopération: Paris, France, 1989.
3. Narjisse, H. Nutrition et production laitière chez le dromadaire. *Options Méditerranéennes-Série Séminaires* **1989**, 2, 163–166.
4. Maskar, U. Bones of the diaphragm in camels. *Acta Anat.* **1957**, *30*, 461–471. [CrossRef] [PubMed]
5. Namshir, N. Regional anatomy of camel internal organs. *J. Agric.* **1982**, *3*, 25.

biology and life sciences forum

MDPI

Proceeding Paper

Principal Anatomy Particularities in Dromedary Compared to Ox: Digestive and Respiratory Systems †

Maya Boukerrou [1,*], Rania Ridouh [1], Faiza Tekkouk-Zemmouchi [1] and Claude Guintard [2]

1 Gestion Santé et Productions Animales Research Laboratory, Institut des Sciences Vétérinaires El-Khroub, Université Frères Mentouri Constantine 1, Constantine 25000, Algeria; anatomiste27@outlook.fr (R.R.); ftekkouk@yahoo.fr (F.T.-Z.)

2 Unité Anatomie Comparée, Ecole Nationale Vétérinaire de l'Alimentation et de l'Agroalimentaire, 44300 Nantes, France; claude.guintard@oniris-nantes.fr

* Correspondence: maya.boukerrou@doc.umc.edu.dz

† Presented at the 10th International Seminar of Veterinary Medicine: Camelids in Algeria & Maghreb, Constantine, Algeria, 20–21 December 2022.

Abstract: The dromedary, also called the one-humped camel, can live and adapt well to the harsh and hot climate of the desert due to its functional anatomical particularities. Both the dromedary and the ox are known as artiodactyls. Anatomically, they have a few similarities but many differences. This paper aims to review the digestive and respiratory anatomy of the two species and enumerate their main particularities. The camel's mouth is thick and unique; it helps them to be selective with regard to food and to avoid any thorny plants. The nostrils of the dromedary are slit-like, with wings that can close to ensure protection against the wind and the sand; they also have a role in the conservation of water. Contrary to the ox, the dromedary's lungs are not lobulated, and the diaphragm has a unique structure. Unlike the bovids, camelids have only three distinct digestive chambers instead of four; there is no clear distinction between the third and fourth chambers. In addition, they have glandular sac areas called "water cells" instead of papillae in the rumen. Their liver is lobulated, and without a gall bladder.

Keywords: comparative anatomy; dromedary; ox; digestive; respiratory

Citation: Boukerrou, M.; Ridouh, R.; Tekkouk-Zemmouchi, F.; Guintard, C. Principal Anatomy Particularities in Dromedary Compared to Ox: Digestive and Respiratory Systems. *Biol. Life Sci. Forum* **2023**, *22*, 13. https://doi.org/10.3390/blsf2023022013

Academic Editors: Amira Leila Dib, Said Boukhechem, Hithem Bougherara and El-Hacene Bererhi

Published: 6 April 2023

1. Introduction

Camelidae includes six species: the Bactrian camel (*Camelus bactrianus*; two humps), the dromedary camel (*Camelus dromedarius*; one hump), lamas, alpacas, guanacos, and vicunas [1]. The dromedary is an animal that is frequently cited as an example of adaptation and resistance to harsh environmental conditions. This is due in part to its anatomy, particularly its digestive and respiratory systems, which are distinct from those of other domestic ruminants. The dromedary and the ox both belong to the order of the artiodactyls, the first to the Tylopoda sub-order and the second to the Ruminantia sub-order [2]. They share several anatomical characteristics (extern and intern anatomy). However, there are some anatomical features specific only to camels. Due to these anatomical particularities, the camel is able to survive in the hot and dry conditions of the desert. The purpose of this paper is to document the main anatomical particularities between these two large ruminants' digestive and respiratory systems.

2. Material and Methods

This paper includes 5 manuscripts: 4 research articles [1,3–5] and one review paper [6]. It focuses on the main differences between the digestive and respiratory systems of the dromedary and the ox.

3. Results and Discussion

The results are shown in Tables 1 and 2.

Table 1. Some particularities of the digestive tract between the dromedary and the ox [1,4,7,8].

	Dromedary	Ox
Mouth	Split upper lip and pendulous lower lip	Upper lip larger than the lower
Dental formula	I1/3 C1/1 PM3/2 M3/3	I0/4 C0/0 PM3/3 M3/3
Soft palate	Extensible and extruded	Not extruded
Salivary glands	Absence of monostomatic sublingual gland	Presence of monostomatic sublingual gland
Esophagus	Enters directly into the rumen	It joints the stomach between the rumen and reticulum
Stomach	3 distinct chambers Presence of glandular sac	4 distinct digestive chambers Absence of the glandular sac
Small intestines	40 m in length, large jejunum	40 m in length, jejuno-ileum
Large intestines	Extremely long (20 m) Blind caecum attached Helical colon	10 m in length, Blind caecum free S-shaped colon with loops
Accessory glands	Liver: lobulated Gall bladder: absent Spleen: not attached to the diaphragm	Liver: not lobulated Gall bladder: Present Spleen: attached to the diaphragm

Table 2. Some particularities of the respiratory system between the dromedary and the ox [3,4,8,9].

	Dromedary	Ox
Nostrils	Slit-like appearance	Separated by a large muffle
Trachea	66–75 incomplete rings of hyaline cartilage	45–55 rings of cartilage with a dorsal crest
Lungs	5 lobes [4] with absence of fissures	7 lobes with presence of fissures
Diaphragm	Ossified [5]	Not ossified

4. Conclusions

The digestive and respiratory systems of the dromedary have some anatomical particularities that allow him to survive in the desert, where food and water are infrequent and the environment is hot and windy [6]. These distinctions aid in our understanding of the camel's way of life and physiological properties.

Author Contributions: Conceptualization, M.B. and R.R.; methodology, M.B.; software, M.B.; validation, R.R., F.T.-Z. and C.G.; formal analysis, M.B.; investigation, M.B.; resources, M.B.; data curation, M.B.; writing—original draft preparation, M.B.; writing—review and editing, M.B., F.T.-Z. and C.G.; visualization, C.G.; supervision, F.T.-Z. All authors have read and agreed to the published version of the manuscript.

Funding: This research received no external funding.

Institutional Review Board Statement: Not applicable.

Informed Consent Statement: Not applicable.

Data Availability Statement: No new data were created or analyzed in this study. Data sharing is not applicable to this article.

Acknowledgments: The authors would like to thank Amira Leila DIB for her invaluable guidance, and Nedjoua LAKHDARA for her assistance with the translation.

Conflicts of Interest: The authors declare no conflict of interest.

References

1. Soliman, M.K. Functional anatomical adaptations of dromedary (Camelus dromedarius) and ecological evolutionary impacts in KSA. In Proceedings of the International Conference on Plant, Marine and Environmental Sciences (PMES-2015), Kuala Lumpur, Malaysia, 1–5 January 2015; pp. 19–22.
2. Berkovitz, B.K.; Shellis, R.P. *The Teeth of Mammalian Vertebrates*; Academic Press: Cambridge, MA, USA, 2018.
3. Bello, A.; Onu, J.E.; Onyeanusi, B.I.; Umaru, M.A.; Suleiman, M.H.; Nzalak, J.O.; Atabo, S.M. Fetal Differentiation of the Lower Respiratory Tract of the One Humped Camel (Camelus Dromedarius): Gross Observation. *J. Hum. Anat.* **2017**, *1*, 000118.
4. Lechner-Doll, M.; Engelhardt, W.V.; Abbas, A.M.; Mousa, H.M. Particularities in forestomach anatomy, physiology and biochemistry of camelids compared to ruminants. *Options Méditerranéennes Ser. B. Etudes Et Rech.* **1995**, *13*, 19–32.
5. Wilson, R.T. The nutritional requirements of camel. In *Seminar on the Digestion, Nutrition and Feeding of Dromedaries*; Tisserand, J.L., Ed.; CIHEAM-IAMZ: Zaragoza, Spain, 1989; pp. 171–179.
6. Chase, M. Camel Anatomy; More Than Just a Hump. *Rev. A J. Undergrad. Stud. Res.* **2019**, *20*, 5.
7. Wilson, R.T. The nutritional requirements of camel. In Proceedings of the Seminar on the digestion, nutrition and feeding of dromedaries, Ouargla, Algerie, 27 February–1 March 1988.
8. Barone, R. *Comparative Anatomy of Domestic Mammals—Volume 3, Splanchnology 1: Digestive System and Respiratory System*; Vigot: Paris, France, 2009. (In France)
9. Ouajd, S.; Kamel, B. Physiological particularities of dromedary (Camelus dromedarius) and experimental implications. *Scand. J. Lab. Anim. Sci.* **2009**, *36*, 19–29.

MDPI
St. Alban-Anlage 66
4052 Basel
Switzerland
Tel. +41 61 683 77 34
Fax +41 61 302 89 18
www.mdpi.com

Biology and Life Sciences Forum Editorial Office
E-mail: blsf@mdpi.com
www.mdpi.com/journal/blsf

www.ingramcontent.com/pod-product-compliance
Lightning Source LLC
Chambersburg PA
CBHW062106220926
51850CB00010B/2917